Global Ocean Governance and Ecological Civilization

This series is a research achievement of frontier discipline and has high-quality publishing resources. The ocean is the most important life support system on earth. Marine ecosystem health is fundamentally important for humankind, and is vital for the development of global economy. The authors of this book series are academicians, renowned experts and scholars, and many front-line researchers who have been working in marine science and technology for a long time. This book series include five volumes (in both Chinese and English), each focusing on a specific topic, include: Living Marine Resources and Biodiversity, Marine Pollution, Green Maritime Operations, Renewable Energy Systems, and Mineral Resource Extraction. It reflects the latest progress of ocean governance and ecological civilization construction, by putting forward original recommendations and solutions for resolving the major cutting-edge scientific and management issues of the ocean. As such, this book series will play a positive role in promoting scientific and technological progress and innovative development in the field of ocean governance and ecological civilization, and have extremely important social value and reference significance.

Hui Liu · John Mimikakis · Fang Sun ·
Jacob P. Kritzer
Editors

Conservation and Sustainable Use of Living Marine Resources and Biodiversity

海洋出版社

中国环境与发展国际合作委员会
China Council for International Cooperation on Environment and Development

Springer

Editors
Hui Liu
Yellow Sea Fisheries Research Institute
Chinese Academy of Fishery Sciences
Qingdao, China

Fang Sun
Environmental Defense Fund
Beijing, China

John Mimikakis
Environmental Defense Fund
Washington, DC, USA

Jacob P. Kritzer
Environmental Defense Fund
Boston, MA, USA

ISSN 2731-460X ISSN 2731-4618 (electronic)
Global Ocean Governance and Ecological Civilization
ISBN 978-981-19-4789-6 ISBN 978-981-19-4790-2 (eBook)
https://doi.org/10.1007/978-981-19-4790-2

Jointly published with China Ocean Press Ltd.
The print edition is not for sale in China (Mainland). Customers from China (Mainland) please order the print book from: China Ocean Press Ltd.

This Springer imprint is published by the registered company Springer Nature Singapore Pte Ltd.
The registered company address is: 152 Beach Road, #21-01/04 Gateway East, Singapore 189721, Singapore

China Council for International Cooperation on Environment and Development (CCICED)
Special Policy Study on Global Ocean Governance and Ecological Civilization

Foreword

The ocean is fundamentally important for humankind. Simply stated, the ocean helps us breathe, regulates our global climate, and through this shapes and defines our living environment and provides for the resources we depend upon. The ocean is also vital for the world's economic development. Three billion people globally rely directly on the ocean for their livelihoods.

However, the oceans and the ecosystem services they provide are under ever more serious threat than before. Global ocean health conditions are being seriously affected by large-scale environmental pressures such as global warming, increased ocean acidification under a continuously higher atmospheric carbon dioxide level, microplastics pollution, and overexploitation of natural resources.

Ocean ecosystem are fragile, highly dynamic, and globally interconnected. There is good cause to manage and govern ocean ecosystems in a manner that in the best possible way ensures a healthy and sustainable ocean supporting prosperous societies now and in the future.

Responding to these challenges and opportunities that the ocean provides us with, the China Council for International Cooperation on the Environment and Development (CCICED) in 2017 initiated a Special Policy Study on *Global Ocean Governance and Ecological Civilization—Building a Sustainable Ocean Economy for China.* The aim was to identify ocean and coastal issues of priority interest to China, to map relevant ongoing national and international ocean initiatives and undertakings, to suggest how China could complement and contribute to these efforts and to recommend areas that could be relevant for China to take leadership on.

The Special Policy Study has chosen six specific ocean related issues to study. An extensive group of Chinese and international experts have been involved in the work.

Considering the key role that the oceans play as the basis for society, humanity, and the global future, the need for more and better governance of ocean uses has been widely recognized. Globally, **Ecosystem-Based Integrated Ocean Management** is accepted as the appropriate approach for ensuring the protection and the sustainable use of coasts and oceans, taking sufficiently into account knowledge and the particularities of the ecosystems to be managed. The Special Policy Study thus considered how China is placed to develop and implement a fully integrated ecosystem-based ocean management system and to take on international leadership in this field.

Living Marine Resources and Biodiversity was another topic considered specifically. Feeding more than nine billion people by 2050 while protecting biodiversity and the natural systems on which life depends is one of the greatest planetary challenges the world faces today. As one of the world's top seafood-producing nations, China has a significant stake in meeting this challenge. While solving them will produce enormous direct benefits for China by securing a large and sustainable supply of high-value seafood and livelihoods in its own waters, it will also create tremendous opportunities for China to demonstrate regional and global leadership among developing countries that lack the capacity to sustain the value of their living marine resources.

There is a tremendous opportunity for China to promote the case for **Marine Pollution** governance. Like most coastal nations, agricultural and industrial requirements for feeding, clothing, and housing the growing population and expanding economies have seriously degraded parts of China's marine and coastal environment. Pollution control is one of the tough battles that the Chinese government must win. Innovation in cleaner production methods and means of reducing discharges and emissions should be actively encouraged to keep pace with the rapid industrial growth.

The Special Policy Study also looked at the issue of **Green Maritime Operations**. As a country trading across all regions around the globe, maritime operation has become a crucial foundation of China's socioeconomic growth. As the dynamic and vulnerable nature of the marine ecosystems has become increasingly understood and appreciated, higher standards for green maritime operations are required both domestically and internationally. While land-sourced pollution is an important concern and contributor to the degradation of the marine environment, China's maritime operation—maritime transportation, port and ship infrastructure, oil and gas exploration, and fisheries—needs to shoulder its share of responsibility in building an ecological civilization and tackling pollution prevention and control.

Achieving the needed renewable energy transition will not only mitigate climate change but also stimulate the economy, improve human welfare, and boost employment worldwide. **Renewable Energy Systems** was considered another topic of particular interest as an emerging sector of the maritime industry. China, the world's biggest energy consumer, is proposing higher green power consumption targets and stepping up its push into renewable energy, also with respect to ocean renewable energy (ORE). Understanding and assessing the environmental impact of ORE installations, operations, and decommissioning is substantially challenging due to, e.g., the lack of baseline environment data and the need for diverse developing technologies. ORE technologies offer opportunities for China to develop a new industry, create jobs, and take advantage of opportunities within its competency to global markets.

Because of the huge reserves found on the deep seabed, **Mineral Resource Extraction** may be of great significance to the world's economic development as well as to the strategic reserve of mineral resources. There are ample opportunities for Chinese industry to engage at all levels of the deep seabed mining value chain, including research, exploration, exploitation, equipment manufacturing, technology design, and mineral processing and to promote deep sea mining as part of the circular economy. In the collaborative work between operators and interested parties, there is a need to focus on reducing the environmental risk as well as sharing of data and experience to ensure the industry adapts best environmental practice and continuous improvement.

Each of the six expert groups presented their findings and policy recommendations in their respective reports. The Special Study Group has synthesized the findings and policy recommendations of the six expert groups reports into one single report, titled "*Global Ocean Governance and Ecological Civilization—Building a Sustainable Ocean Economy for China*", accessible from the website of CCICED. (http://en.cciced.net/POLICY/rr/prr/2020/202009/P020200916750698907787.pdf).

With the intention to reach a wider audience, the Special Policy Study decided to publish the *Global Ocean Governance and Ecological Civilization* Book Series that you now have in your hands. This book series includes five volumes, which are expanded version of the respective reports of the five Groups, namely, the **Living Marine Resources and Biodiversity**, **Marine Pollution**, **Green Maritime Operations**, **Renewable Energy Systems**, and **Mineral Resource Extraction**. We hope that students, researchers, managers, businesses and interested citizens can make use of the facts, analyses and wisdom that we have collected.

Hanghzou, China
Tromso, Norway

Su Jilan
Team Leader of CCICED SPS on
Global Ocean Governance
and Ecological Civilization
Honorary President of the Second
Institute of Oceanology
Ministry of Natural Resources
Academician of Chinese Academy
of Sciences

Jan-Gunnar Winther
Team Leader of CCICED SPS on
Global Ocean Governance
and Ecological Civilization
Specialist Director
Norwegian Polar Institute

Preface

Maintaining global food security and marine economic development, while protecting marine biodiversity and the natural ecosystem on which life depends, is one of the biggest challenges facing coastal countries throughout the world. As one of the world's top seafood producing nations, China has a significant stake in solving this challenge. China's marine fisheries (including catch fishery and marine aquaculture, i.e. mariculture) have made important contributions towards satisfying the demand for high-quality aquatic food at home and abroad, promoting rural economic prosperity and increasing fishermen's income, optimizing the national dietary structure and ensuring food security. However, China faces major, unprecedented challenges surrounding the governance of living marine resources and marine ecosystems. Large-scale habitat degradation, a decline in most fishery resources, and the environmental impact and spatial limitation for mariculture are among the threats to the country's sustainable production capacity.

The Chinese government and academia have recognized these problems and taken measures to actively respond, including ecological restoration and enhancement release aimed at restoring living marine resources and their habitats; the seasonal moratoria and "dual control on fishing vessel and power" to combat overfishing, and the ongoing pilot trials of quota control of marine capture fisheries; the zoning of permitted, restricted and prohibited aquaculture areas (here, aquaculture refers to both freshwater and marine aquaculture), and strengthening of waste water treatment for enhancing aquaculture governance; as well as extensive scientific and technological research to implement and improve the above management measures. However, in terms of actual results, China's governance of living marine resources, and relevant scientific research and development, are far from enough. Given the huge population, extensive coastline and continued development of the marine economy, the path towards a marine ecological civilization is not an easy one.

Still, China has demonstrated strong political will and steady progress in the management of its living marine resources. Building off an existing Special Policy Study on Global Ocean Governance and Ecological Civilization (commissioned by the China Council for International Cooperation on Environment and Development

(CCICED)), this monograph presents a global landscape analysis of China's management and utilization of living marine resources. The book closely examines current challenges and existing management approaches, and addresses future challenges and long-term vision. Given the increasing threat of climate change, there is a chapter dedicated to examining climate impact on marine species. After focused study and presentation of key international case studies from around the world, the report then proposes several recommendations to support China's next steps as it seeks to create an ecological civilization, both domestically and abroad, suitable for living marine resources.

The book includes eight chapters. Chapter 1 presents background research, introducing the value of marine biological resources and biodiversity from the perspective of important life support systems, and discusses the current status of the development and utilization of marine biological resources in China and around the world. The main content is then presented in two parts: Part I (Chaps. 2–5) addresses the status quo, trends and challenges of China's living marine resource governance; Part II (Chaps. 6–8) surrounds development trends and goals for living marine resource governance.

Chapter 2 introduces China's mariculture industry and its governance, highlighting its significance as an important source of seafood by complimenting/replacing capture fisheries, and discusses the challenges in terms of social economy, science and technology. Chapter 3 introduces capture fisheries in China, covering scale and characteristics, development history and prominent issues, and highlighting the continuous improvement of China's fishery governance system as well as urgent issues to be solved. Chapter 4 presents an in-depth discussion of the current status of living marine resource management from the perspective of habitat and biodiversity conservation, and the relevant social, economic and environmental challenges. Chapter 5 focuses on the impact of climate change and discusses the environmental challenges faced by living marine resources from the perspective of global change. Chapter 6 reviews the policy and management system for China's marine living resources, emphasizing the development trend and existing institutional mechanisms of marine living resources management. Chapter 7 presents international experiences in the management of living marine resources, which have been selected to provide ideas and inspiration that match relevant needs and opportunities in China. The chapter is broken down into four sections, focusing on experiences with strengthening monitoring; holistic and integrated environmental planning; promoting long-term value over volume in wild fisheries; and scaling and adapting to proactively manage in a changing world. Finally, Chap. 8 makes the connection between China's unique challenges and possible solutions from a global perspective, and presents a set of actionable policy recommendations for China to achieve the ambitions of ecological civilization.

The authorship of this monograph include more than 30 experts from research institutes and higher education in China and NGOs in the US. The names of main authors are given in relevant chapters.

This book can be used as a reference for professional and technical personnel engaged in research, management or policy related to living marine resources, both in China and abroad. The structure of the book and model for our research and presentation of policy recommendations can also be utilized in other topic areas or countries to help guide public policy. The book can be also used as a reference for teachers and students at colleges and universities.

Qingdao, China — Hui Liu
Washington, DC, USA — John Mimikakis
Beijing, China — Fang Sun
Boston, MA, USA — Jacob P. Kritzer

Acknowledgements

We would like to thank the China Council for International Cooperation on Environment and Development (CCICED) of the Ministry of Ecology and Environment of People's Republic of China (MEE) for convening and funding the special policy study on Global Ocean Governance and Ecological Civilization.

We appreciate the helpful and constructive comments from Dr. Su Jilan, Dr. Jangunnar Winther, Dr. Art Hansen and Dr. Liu Shijin, which greatly improved this report.

We would give special thanks to Mr. Li Yonghong, Mr. Zhang Huiyong, Ms. Yao Ying, Mr. Li Panwen and Ms. Fei Chengbo from CCICED, and Ms. Wang Minfang from the Second Institute of Oceanology, Ministry of Natural Resources, for your tireless work to ensure the success of this project.

We also would like to thank Dr. Lisa Svensson from the United Nations Environment, Mr. Nishan Degranian from the World Economic Forum, Dr. Dai Minhan from Xiamen University, Dr. Wang Juying from the National Marine Environmental Monitoring Centre of MEE, Dr. Sun Song from the Institute of Oceanology of Chinese Academy of Sciences, Dr. Han Baoxin from the South China Institute of Environmental Sciences of MEE and Dr. Birgit Njaastad from Norwegian Polar Institute, for your valuable support of this work.

Editorial Board

List of Editorial Committee for the Book Series—Global Ocean Governance and Ecological Civilization

Executive Summary

Feeding more than nine billion people by 2050, while protecting biodiversity and the natural systems on which life depends, is one of today's greatest planetary challenges. As one of the world's top seafood producing nations, China has a significant stake in solving this challenge.

The ocean shelters a vast array of species, produces most of the world's oxygen and nourishes billions of people who depend upon the sea for food and livelihoods. However, while the ocean's ability to produce food is enormous, it is limited. About one-third of fisheries for which data exist are overexploited or collapsed; the actual number may be even higher, as many countries do not have capacity to measure their fisheries' output, much less manage them.

Aquaculture production has nearly surpassed wild fisheries as a source of food, and there is considerable potential in the ocean for even greater production. Still, aquaculture can have negative impacts: it can displace native coastal and marine ecosystems, require large inputs of wild fish for feed, introduce non-native species and diseases, and cause significant pollution. Sustainably managing living marine resources to optimize economic production over the long term—while minimizing damage to the ecosystem—is not an easy task, although there are many proven solutions and new ones in development to achieve these goals.

Unfortunately, climate change will likely increase the challenge of feeding people sustainably. Warming and acidifying waters are altering the productivity of many marine species and driving others across borders, intensifying the struggle for resources among competing countries. More extreme storms, altered weather patterns and disrupted water and nutrient cycles will increasingly stress coastal food production systems. To date, few countries have found ways to handle these changes effectively at scale.

China faces similar challenges in managing its living marine resources to those of other countries; however, the scale of China's economy makes its situation more extreme. Rapid economic growth along China's coasts over the last 40 years has imposed significant costs on the coastal and marine environment. Development, mariculture and pollution have damaged more than half of the country's coastal wetlands, nearly 60% of mangroves, 80% of coral reefs, and significant portions of the seagrass beds, tidal marshes and tidal flats that once provided a critical habitat for a diverse array of marine life. China produces enormous quantities of wild and farmed seafood—by far, it is the world's top producer by volume—but the rate of extraction and exploitation have outstripped the ability of marine ecosystems to absorb the impact, and top marine predators have been almost entirely eliminated. In addition, China employs more people in the fishing and aquaculture sector than does any other nation, making the social aspect of bringing these industries under control much more challenging.

Despite the difficulty, China has begun making significant progress in addressing these challenges. President Xi Jinping has committed to creating an ecological civilization, a balanced integration of economic and environmental goals that is now enshrined in the Constitution. And the government has followed with bold, concrete actions, including strengthening the seasonal closings of nearly all domestic fisheries each year, protecting habitat and establishing marine reserves, imposing stricter standards for discharging pollutants into the sea, shutting down illegal mariculture operations and establishing a total acreage limit for mariculture.

However, more must be done if China is to reestablish healthy coastal and marine ecosystems that can sustainably provide the levels of nutritional and economic benefits the country has historically enjoyed. To create an ecological civilization, legal protections for living marine resources must be strengthened, monitoring expanded and compliance improved, and more critical habitat restored and protected. Furthermore, because climate change is affecting living marine resources on a global scale, and because so many of these resources are shared, especially within Asia, an ecological civilization at home depends upon stronger regional and global governance to ensure living marine resources are managed sustainably at scale.

Although these challenges are not easy, solving them will produce enormous direct benefits for China by securing a large and sustainable supply of high-value seafood and livelihoods in its own waters. It will also create tremendous opportunities for China to demonstrate regional and global leadership on oceans. President Xi's Maritime Silk Road initiative represents an historic opportunity for China to promote global marine governance and advance the UN Sustainable Development Goals (SDGs). Through collaboration with countries along the Silk Road, China can promote the conservation of marine biodiversity, empower women and small-scale fishing communities to develop their marine resources sustainably and feed themselves, and create a platform for countries to work together to manage the impacts of climate change on living marine resources.

Building off an existing Special Policy Study on Global Ocean Governance and Ecological Civilization (commissioned by the China Council for International Cooperation on Environment and Development (CCICED)), this monograph presents a global landscape analysis of China's management and utilization of living marine resources. The book closely examines current challenges and existing management approaches, and addresses future challenges and long-term vision. Given the increasing threat of climate change, there is a chapter dedicated to examining climate impact on marine species. After focused study and presentation of key international case studies from around the world, the report then proposes several recommendations to support China's next steps as it seeks to create an ecological civilization, both domestically and abroad, suitable for living marine resources.

This report proposes several recommendations to guide China's next steps as it endeavors to create an ecological civilization domestically and abroad for living marine resources, including:

1. Strengthen legal protections for coastal and marine ecosystems, while promoting sustainable production.
2. Implement a high-tech monitoring system to combat corruption and illegal fishery operations.
3. Develop mariculture sustainably under the framework of ecosystems.
4. Upgrade technologies continuously, to improve quality and efficiency of mariculture.
5. Restore marine ecosystem functions to improve resilience to global change.
6. Maritime Silk Road networking to promote sustainable ocean governance.
7. Carry out a national plan for implementation of the UN 2030 Agenda for Sustainable Development Goals.
8. Continuously promoting gender equity in fisheries sector.

Contents

Contributors

Robert Boenish Environmental Defense Fund, Washington, DC, USA

Ling Cao Xiamen University, Xiamen, China

Shuo Cheng Xiamen University, Xiamen, China

Yang Han Development Research Center of the State Council, Beijing, China

Ziyu Jiang Shanghai Jiao Tong University, Shanghai, China

Jacob P. Kritzer Environmental Defense Fund, Boston, USA

Chunyu Liu Shanghai Jiao Tong University, Shanghai, China

Hui Liu Yellow Sea Fisheries Research Institute, Chinese Academy of Fishery Sciences, Qingdao, China

John Mimikakis Environmental Defense Fund, Washington, DC, USA

Jilan Su Second Institute of Oceanology, Ministry of Natural Resources, Hangzhou, China

Fang Sun Environmental Defense Fund, Beijing, China

John Virdin Environmental Defense Fund, Durham, USA;
Duke University, Durham, USA

Zihan Wang Shanghai Jiao Tong University, Shanghai, China

Daniel Willard Environmental Defense Fund, Austin, USA

Jeff Young Environmental Defense Fund, San Francisco, USA

Pengyan Zhang Yellow Sea Fisheries Research Institute, Chinese Academy of Fishery Sciences, Qingdao, China

Chapter 1
Overview of Living Marine Resources and Biodiversity

Hui Liu, Jacob P. Kritzer, Fang Sun, and Pengyan Zhang

The planetary challenge the world faces—namely, how to feed a growing population without destroying the ecosystem that supports us—is unquestionably not China's alone. However, given the enormity of seafood it produces and consumes, the size of its marine economy and the multitudes it employs in its seafood industries, perhaps few other countries have as much at stake in ensuring the planet's marine resources are sustainable. Likewise, few others are as influential regionally and globally and have as much to offer diplomatically, technically and scientifically. As China seeks to create an ecological civilization, there is both an opportunity and a need to join with other countries to protect the ocean that nourishes and sustains us all.

1.1 The Value of Living Marine Resources

Ocean ecosystems harbor immense biodiversity, an abundance of life that provides numerous benefits. These living marine resources have cultural value in many places, playing a central role in mythology, religion, symbols and stories worldwide, from

H. Liu (✉) · P. Zhang
Yellow Sea Fisheries Research Institute, Chinese Academy of Fishery Sciences, Qingdao, China
e-mail: liuhui@ysfri.ac.cn

P. Zhang
e-mail: zhangpy@ysfri.ac.cn

J. P. Kritzer
Environmental Defense Fund, Boston, USA
e-mail: jake@neracoos.org

F. Sun
Environmental Defense Fund, Beijing, China
e-mail: sunfang@cet.net.cn

H. Liu et al. (eds.), *Conservation and Sustainable Use of Living Marine Resources and Biodiversity*, Global Ocean Governance and Ecological Civilization,
https://doi.org/10.1007/978-981-19-4790-2_1

antiquity through to modern times. Many people enjoy directly interacting with ocean life through scuba diving, birdwatching, sport fishing and other forms of recreation, all of which support a lucrative global tourism industry. Living marine resources also provide important regulating and supporting services, such as oxygen production, nutrient cycling, water filtration, carbon sequestration and storm buffering. The value of these services to a growing ocean economy is difficult to estimate, but is likely on the order of hundreds of billions, if not trillions, of U.S. dollars worldwide each year (de Groot et al., 2012). Yet, the full breadth of services provided by living marine resources is not well appreciated. Consequently, too many policies fail to safeguard living marine resources; biodiversity is in decline and ecosystem services and the value they provide are being lost, posing a risk to human health, food security, welfare and livelihoods.

Perhaps the most significant and widespread ecosystem service delivered by marine life is provision of seafood through wild capture fisheries and mariculture. In 2018, global fisheries product was estimated to reach a new record of 179 million tonnes (Hereinafter referred to as MT), including 156 MT used for human consumption, equivalent to an estimated annual supply of 20.5 kg/capita (FAO, 2020). For some, seafood makes for a more interesting and varied diet and provides a deeper connection to the sea. For others, especially in many parts of the developing tropics, seafood is a critical component of food and nutrition security. Indeed, 20% of the world's population is critically dependent upon seafood as a source of micronutrients, so called because only small amounts are required to fulfill vital physiological functions (Golden, 2016). Fisheries and aquaculture not only provide food, but income and livelihood, to 18.28 million fisheries practitioners in China (BOF, 2020). These benefits are not limited solely to those who harvest food from the sea, but also extend to the many support services—i.e. gear manufacturers, vessel mechanics, etc.—as well as processors, restaurants and other businesses along the supply chain that are also built on seafood. The value of these other seafood-dependent industries can substantially add to the employment and revenue generated by the production sectors.

These promising findings notwithstanding, most future increases in the global supply of seafood will come from mariculture production. In the Agricultural Outlook 2017–2026 (OECD-FAO, 2017), it is predicted that by 2026, world fisheries production will reach 194 MT, including 102 MT of aquaculture products. And according to FAO (2020), world fish products will see a 15% increase and reach 204 MT in 2030 (Fig. 1.1), with the percentage contribution from aquaculture also to increase from the current 46%. This is about half of the increase seen over the past 10 years; in 2030, per capita fish consumption is estimated to be 21.5 kg.

The volume of aquatic food generated by aquaculture has long been on the rise, and likely will soon equal and surpass production from wild capture fisheries. Despite considerable potential for additional mariculture production worldwide (Gentry et al., 2017), the growth rate of world aquaculture has slowed over the last 40 years, and will probably see further decrease over the next decade (Fig. 1.2) (FAO, 2020). One possible reason is that aquaculture is a resource-dependent industry, and would be limited by many factors including space, feed source, energy and labor costs, etc.

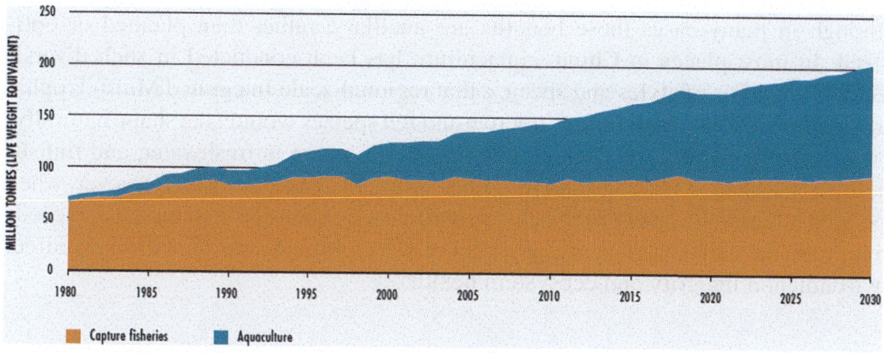

Fig. 1.1 World Capture Fisheries (orange) and Aquaculture (blue) production in 1980–2018 and estimated production for 2020–2030 (*Source* FAO, 2020)

On the other hand, aquaculture can pose considerable risks to marine ecosystems and other value generated by living marine resources (including wild capture fisheries) through unsustainable capture of wild seedlings, such as eel *Anguilla japonica*; through excessive use of trash fish to feed farmed fish (Cao et al., 2015); and through encroachment upon key habitats for wild species, thus generating environmental pollution and triggering opportunistic species (Liu & Su, 2017). For example, large-scale infestations of green algae *Ulva prolifera* off the coast of the northern Yellow Sea have occurred repeatedly in recent years, threatening the valuable tourism industry in the region. While multiple factors contributed to these infestations, the most important driver has been identified as coastal seaweed farms from as far away as Jiangsu Province that provided substantial additional surface area for algal growth (Pang et al., 2010; Liu et al., 2013; Sun et al., 2018).

At the same time, extractive mariculture species, such as seaweed and bivalves, can also generate ecological benefits, including water filtration and nutrient extraction, and creation of shelter and nursery habitats for juveniles (Liu et al., 2017),

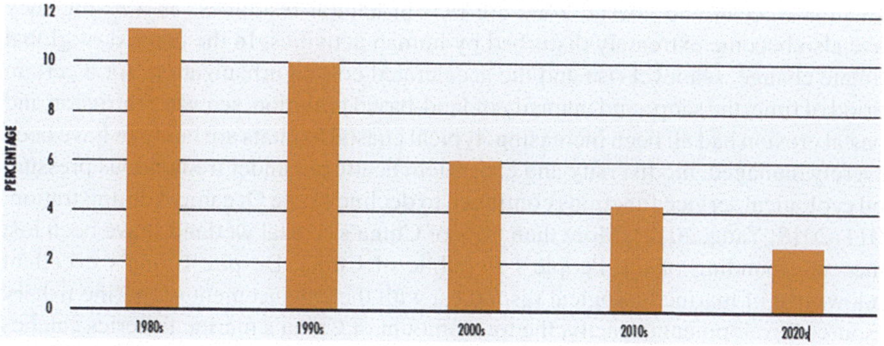

Fig. 1.2 Annual growth rate of World Aquaculture, 1980–2030 (*Source* FAO, 2020)

although in many cases these benefits are ancillary, rather than planned or optimized. In most places in China, aquaculture has been conducted in such diversified combinations of styles and species, that regional-scale Integrated Multi-Trophic Aquaculture (IMTA) with both extractive and fed species would take shape naturally; good examples of this are rice-fish integrated cultivation in freshwater, and finfish-shellfish-seaweed in net-cage and longline culture in coastal oceans. However, when operated intensively at very high density, particularly when cultured biomass exceeds environmental carrying capacity, any kind of aquaculture operation will likely affect environmental integrity and ecosystem health.

1.2 China's Living Marine Resources

The exclusive economic zone (EEZ) of China covers a wide coastal ocean area from a coastline that spans 18,000 km, across more than 44° of latitude and 32° of longitude. This EEZ, with a jurisdiction of approximately three million km^2, stretches from the tropical waters of the Beibu Gulf and Zengmu Shoal (also known as James Shoal) in the South China Sea, through the sub-tropical East China Sea, and into the temperate Yellow Sea in the north. Together with the Bohai Sea, it forms three large marine ecosystems (LMEs). The superior natural environment of China's oceans provides extremely favorable living conditions for marine life. Numerous spawning grounds, nursery grounds, feeding grounds and wintering grounds have been formed here, giving birth to tens of thousands of marine life species. Based on publication between 1922 and 1992, there were 22,561 species of living marine resources in the China seas (Huang, 2008); according to an updated statistic, the number has increased to over 28,000, comprising 13% of the global total (Huang & Lin, 2012). Among these, there are more than 2500 species of economically valuable finfish, 685 species of crabs, 90 species of shrimps, and 84 species of cephalopods. There are also more than 300 species of important fishery resource organisms recorded, among which are 60–70 common economic fishery species of particularly high yield (Tang, 2012).

China's ocean and coastal zone are rich in natural resources; as a result, they have also become extremely disturbed by human activities. In the context of global climate change, sea level rise and the accelerated coastal urbanization, for a certain period of time, the scope and intensity of land-based pollution, seawater intrusion and coastal erosion had all been increasing; typical coastal habitats are being or have been severely damaged, biodiversity and ecosystem health are under tremendous pressure and ecological service functions continued to decline (State Oceanic Administration, 2011–2018; Yang, 2017). More than 50% of China's coastal wetlands have been lost since the founding of the People's Republic of China. Despite China's excellent endowment of marine biological resources, with the enhancement of marine fishery resource development capacity, the total amount of China's marine fisheries catches has increased year to year in the last decade. Due to the limited reproduction and regeneration capacity of living marine resources, China's coastal waters have shown overfishing and fishery resource decline, and deterioration of the coastal ecological environment.

Although it was estimated that the total allowable catch for China's offshore fishery resources was about nine MT (Han, 2018), China's marine fisheries output reached 9.41 MT in 1995, indicating that the utilization of living marine resources had already reached the full extent. However, China's offshore fishing continued to grow after 1995, reaching a historical peak of 14.08 MT in 1999, and so far has remained at a high level of more than 10 MT. Over-exploitation and utilization for 25 consecutive years has led first to serious decline in a number of high-value fishery resources among the top predators of the marine food chain (such as red sea bream *Paragus major*, Japanese flounder *Paralichthys olivaceus*, large yellow croaker *Larimichthys crocea*, etc.), followed by overexploitation of species at the middle level of the food chain (such as *Thamnaconus modestus*, Pacific herring *Clupea pallasii*, small yellow croaker *Larimichthys poyactis*); at the same time, in order to produce fish meal, the fishing pressure on lower trophic species, such as Japanese anchovy *Engraulis japonicus*, has continued to increase. In fact, the long-term fishery statistics can tell the difference. For example, the top three catch species in 1970 were largehead hairtail *Trichiurus lepturus*, Japanese scad *Decapterus maruadsi* and large yellow croaker, and the total output of the three accounted for 60% of the marine catch in that year. However, in 2019, the top three catch species were hairtail, anchovy and swimming crab *Portunus trituberculatus*, whose total output accounted for only 20% of the marine catch, and even the sum of the top 10 catch species was less than 46% of the total catch that year (BOF, 1971/2020). This turn from selective overfishing to non-selective overfishing has completely changed the structure of China's living marine resources. Most of the high-value fish stocks have been fully exploited or suffer from overexploitation, and some high-value species are even on the verge of extinction.

Overfishing has reduced China's ability to produce high-value seafood and rendered its ecosystems more vulnerable to human impact and climate change. Indeed, large-scale coastal development and land reclamation has posed additional stress to the fragile populations of marine living resource, and further weakened their ability to recruit and rebuild. Enhancement release has been conducted in China for about two decades. In recent years, China spent around one billion CNY annually on enhancement release to rebuild important fish stocks; more than 200 marine and freshwater species are released into natural bodies of water each year. During the 12th Five-Year Plan period, China invested nearly five billion CNY and released more than 160 billion seedlings of aquatic organisms. Nevertheless, marine fishery resources have not been significantly restored.

At the same time, China has increasingly relied on distant water fisheries and imports to meet the domestic market demand for high-end aquatic products (Ministry of Agriculture and Rural Affairs, MARA, 2018). However, fisheries in distant waters seldom offer significant potential for additional supply, and often lack the necessary investments for sustainability (e.g., technical expertise, technology, financing). Similarly, overseas sources of capture and mariculture products will need to be developed sustainably in order to ensure a reliable long-term supply. If China is to meet the growing demand for high-value seafood that its increasingly affluent population prefers, it will need to restore its coastal marine ecosystems, make domestic seafood production sustainable and find ways to encourage other countries to manage their resources sustainably, all while building resilience to climate change. These changes can protect and nurture the diverse industries beyond the production sectors that depend on seafood in China.

Indeed, the ocean economy, including the value generated by living marine resources and other aspects of marine ecosystems, has become an important part of China's economic growth. In 2019, the value of main national marine product totaled 35,724 billion CNY, an increase of 7.5% over the previous year. The marine product value accounted for 3.6% of the domestic GDP and 6.8% of the GDP of all the coastal provinces (MNR, 2020). Marine living resource is the most important contributor to fishery economy, which ranks third in value in ocean economy, with an added value of 471.5 billion CNY, comprising 13.2% of the total added value of ocean economy (3572.4 billion CNY).

Marine fisheries, including capture fisheries and aquaculture, and related industries are essential to China's marine economy. In 2019, the total value of China's fishery economy was 2.6 trillion CNY, of which the output value of fisheries was 1.29 trillion CNY, and the related industry and construction, communication and services' output value was about 1.3 trillion CNY. The largest proportion of fishery output value was freshwater aquaculture (618.7 billion CNY), followed by mariculture (357.5 billion CNY) and marine capture fisheries (211.6 billion CNY). Recreational fishery accounted for 96.4 billion CNY, showing much lower annual growth rate (6%) compared to the previous year (18%), despite the continued increase. There were about 9,323 aquatic product processing enterprises in the country, and the annual processing capacity was about 21.71 MT, including 17.76 MT of processing capacity for seafood (Fig. 1.3).

1.3 Linkages Between Living Marine Resource Management and Marine Biodiversity

In China, as in most other countries, management of living marine resources including conservation of marine habitats and biodiversity is a policy priority. However, habitat and biodiversity conservation affect, and are affected by, production of seafood through wild capture fisheries and aquaculture. Whereas healthy habitats and high biodiversity can enhance aquaculture and fisheries production, seafood production typically comes at some cost to habitat and biodiversity (Willison & Côté, 2009). While this means that these sectors cannot always coexist, some win–win solutions are in place across the globe to simultaneously promote fisheries, aquaculture and biodiversity. Indeed, the long-term convergence in governance of biodiversity and fisheries has arguably been the most important ocean policy development over the past three decades, and management of aquaculture is increasingly being integrated.

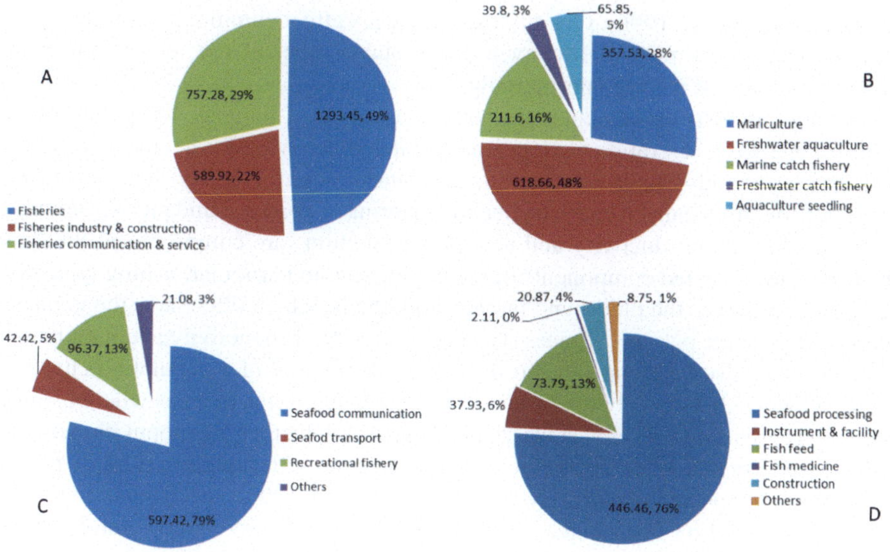

Fig. 1.3 Structure and composition of China's Fishery Economy, 2019. **a** Value and percentage of major components for China fishery economy; **b** Value and percentage of major components for China fishery; **c** Value and percentage of major components for China fishery-related industry and construction; **d** Value and percentage of major components for China fishery-related communication and service (*Source* BOF, 2020)

1.3.1 Effects of Fishing on Biodiversity

Fishing can compromise marine biodiversity in several ways, both through direct extraction and ancillary effects of fishing (Willison & Côté, 2009). Although it is difficult to drive many marine species to biological extinction—due to their broad geographic ranges, highly productive life history strategies and extensive dispersal capabilities—this is not true of all species. Marine mammals, sea turtles, and many large, long-lived and slow-growing fishes and invertebrates do not exhibit these trends; consequently, species in these groups have faced or are at risk of extinction. These species represent very unique components of marine biodiversity that play important ecological functions, so their loss has broader ecological significance.

In fact, even if overfishing does not result in outright biological extinction of a species, populations can become depleted to the point of ecological extinction, whereby important functions of a species in the ecosystem are effectively lost due to low abundance. The loss of these functions can have flow-on effects for other species, with some benefiting and some suffering, depending upon the nature of their ecological interactions with the depleted species. This could lead to additional depletion of species that depend upon the overharvested species as prey or through positive symbiotic relationships, and increased abundance of other species that are prey, competitors, or have adverse symbiotic relationships with the overharvested species (Christensen, 1996). Overall, these changes affect community composition, an important biodiversity metric, even if the total number of species present in an ecosystem does not change substantially.

Fishing can also fundamentally alter the structure of harvested populations in terms of age, size, sex ratio, spatial distribution and behavioral traits through selective harvest of individuals with certain attributes. When such restructuring takes place, a population can appear to be healthy in terms of coarse abundance or biomass metrics, while its productivity and ecosystem functions are compromised through the loss of the selected components of the population. In particular, fishing typically selectively removes the oldest and largest individuals, which often have the greatest reproductive value per unit of mass (Hsieh et al., 2009). Productivity and population resilience can also be compromised if the spatial structure of a population changes through loss of distinct subpopulations, and spawning groups in areas where fishing pressure is greatest (Berkeley et al., 2011). Loss of life history and population diversity are important considerations that often receive too little attention if biodiversity is measured by species numbers alone (Rice & Gislason, 1996). Often the individuals typically selected by fishing gears have very different dietary and migratory characteristics than those of younger and smaller individuals, which affects biodiversity through trophic interactions and movements.

It is possible that diminished components of a population can be recovered through alleviation of overall fishing pressure or changes in selectivity of the gear. This can improve population and life history diversity, as well as broader ecological structure, through the recovery of lost ecosystem functions. However, a wide array of research over the past few decades illustrates that growth, longevity and behavioral characteristics can have a strong genetic basis; fishing can not only have near-term effects on demographic structure, but long-term effects on genetic structure (Hutchings & Reynolds, 2004; Van Wijk et al., 2014). Loss of genetic diversity can be much slower to reverse than excessive mortality, especially if natural selection and genetic drift work against the spread of rare genes (Hutchings & Reynolds, 2004).

1.3.2 Contributions of Biodiversity to Fisheries

Adverse effects of fishing on biodiversity can be detrimental to fisheries, as they rely on biodiversity for their success. Having access to a wide range of species in an ecosystem allows fishing businesses and communities more avenues for prosperity, despite declines in the abundance of certain species, whether due to natural factors or overharvest. Highly specialized fisheries that target one, or a small number of, species can be efficient and profitable—especially if the species have high market value—but are also much more vulnerable as a result of that dependence. Diversified fishing portfolios can provide economic resilience, while at the same time providing ecological benefits from spreading impact more evenly across the ecosystem. Such a risk-spreading approach is the rationale for the "balanced harvest" fishing strategy.

Even if a fishery does not rely directly on other species in an ecosystem for harvest, those species can still support the fishery through a number of ecosystem functions. Conserving the biodiversity of an ecosystem means conserving those species that form habitat, serve as prey, control parasites, produce oxygen, process nutrients, provide water filtration and serve other ecological functions (Diana, 2009). Although most traditional stock assessment models do not account for these functions, collectively they shape the growth, mortality and reproductive characteristics of target species for a fishery. This is not to say that all species in an ecosystem make positive contributions to the productivity of key species, for predators and competitors adversely affect key demographic rates. Nevertheless, the life history strategy of any given species has evolved to optimize reproductive success within the bounded ecological context of its ecosystem, and that context is defined in large part by the constituent biodiversity.

The importance of biodiversity in shaping the functioning of the ecosystems that support fishery species is even greater when considering resilience to environmental perturbations. Ecological niches left unfilled through depletion or loss of key species create an opening for the establishment of invasive species, whereas rich biodiversity produces more ecological and genetic assets in the ecosystem that can counter invasive species. Climate change presents even greater environmental change than invasive species, and indeed climate change can open ecological space in which invasive species can take hold. However, conserving biodiversity gives ecosystems a better chance of adapting to climate impacts by providing more biological resources for adaptation. Conserving and capitalizing on the fundamental structure and function that has enabled marine ecosystems to support species important to fisheries is the far more prudent and precautionary tact.

1.3.3 Interactions Between Aquaculture and Biodiversity

Aquaculture can impact biodiversity through the inputs required for operation, such as seedlings and sources of feed taken from the marine environment, and requirement of physical space and resulting displacement of habitat; and also through its outputs, which can include nutrient loading and pollution, and spread of disease or escapement of contained species into the wild.

Instances in China in which aquaculture operations rely on wild seedlings, such as the culturing of eel *Anguilla japonica*, as well as the harvesting and conversion of lower trophic level species or juvenile fish into aquaculture feed, will have similar impacts on biodiversity as those mentioned in Sect. 1.3.1 (Cao et al., 2015). Creation of aquaculture facilities often means displacement of natural habitats (Diana, 2009; Liu & Su, 2017). Fisheries certainly impact habitats as well, but there is potential for recovery of the habitat after contact with the gear. Many fishery management strategies aim to minimize impacts through gear and effort restrictions, and to explicitly protect habitats or allow for their recovery through spatial and temporal controls. In contrast, habitat effects of aquaculture are often permanent. However, certain types of aquaculture, specifically shellfish and seaweed, can replace functions that have been lost when other habitats are lost (Neori et al., 2004). These functions can include shelter for fishes and invertebrates, water filtration, oxygenation, nutrient processing and carbon sequestration (Chopin et al., 2001).

The impacts of aquaculture on biodiversity are not limited to habitat loss. Aquaculture operations can produce a variety of wastes that increases the pollution load in an ecosystem, to the detriment of habitats away from the facility and other species (Eng et al., 1989). The high densities in which organisms are often grown for aquaculture production can promote disease outbreaks that can spread beyond the facilities (Murray & Peeler, 2005). These organisms can themselves escape, which can introduce non-native species into an ecosystem. Even if the species being grown by an aquaculture operation is native to the area, the individuals are often genetically inferior for not having endured the rigors of natural selection (Diana, 2009). This means that they can dilute the natural gene pool if they escape and interbreed with wild individuals, compromising genetic diversity.

Aquaculture is also different from wild fisheries in that production is less dependent on the characteristics of the surrounding ecosystem. Whereas fishing aims to capitalize on the natural production that results from a wide range of interacting species and processes, aquaculture aims to create and control as many of those factors as possible. This means shaping the timing and location of the life cycle, providing food and shelter from predators, enhancing disease resistance, selecting or establishing optimal temperature and other conditions, and so on. Nevertheless, aquaculture that takes place within natural coastal and marine systems will be affected by the local water quality, pathogen load and other factors, all of which are shaped to some degree by local biodiversity and the ecosystem functions it provides. While this dependence might not be as strong as that of wild fisheries, nevertheless the linkage exists.

1.4 Aquatic Food Supply Worldwide and in China

Out of all the services provided by living marine resources, human beings depend most heavily on the production of aquatic products. In 2018, total world fisheries and aquaculture was 178.5 MT in volume, of which 88% (156 MT) was utilized for direct human consumption. The remaining 12% (22 MT) was used for non-food purposes, of which 82% (or 18 MT) was used to produce fishmeal and fish oil (FAO, 2020).

1.4.1 Aquatic Food Consumption as a Means to Reduce Hunger and Food Insecurity

Apparent per capita fish consumption continued its rising trend and reached the new height of 20.5 kg in 2018 (FAO, 2020), which accounted for 17% of the global population's intake of animal proteins and 7% of all proteins consumed. As the social and economic development is still imbalanced globally, moderate or severe food insecurity still affects 1/4 of the world population, and has been increasing over the past six years. Over half of the population in Africa, almost one-third in Latin America and the Caribbean and more than one-fifth in Asia are food insecure (FAO, IFAD, UNICEF, WFP & WHO, 2020). Utilization of marine living resources has provided vital means of reducing hunger and promoting social equity for 3.3 billion people, with 20% of their average per capita intake of animal proteins sourced from fishery products. As a matter of fact, per capita fish consumption does not vary so much between developing countries (19.4 kg) and the world average (24.4 kg, 2017). While in countries such as Bangladesh, Cambodia, Ghana, Sri Lanka and several Small Island Developing States (SIDS), average per capita intake of fish-source animal proteins may reach 50% or more (FAO, 2020).

1.4.2 Aquatic Product Trade as a Means to Mitigate Social and Economic Inequity

Besides the provision of food and food security to most of the population worldwide, fisheries also generate economic opportunities through trade, which can provide new market opportunities for fishing communities. With the proper social and governance structures in place that integrate inclusion and equity, fisheries production and trade can potentially help to mitigate social inequality and food insecurity. While developed countries still dominate fish imports in terms of volume and value, urbanization and expansion of the fish-consuming middle class have also fueled demand growth in developing markets, outpacing that of developed nations. Imports of fish and fish products of developing countries represented 31% of the global total by value, and 49% in quantity in 2018, compared with 12 and 19%, respectively, in 1976 (FAO, 2020). Though Europe and North America are characterized by a fish trade deficit, Africa has become a net importer in volume terms, but a net exporter in terms of value. African fish imports, mainly affordable small pelagics and tilapia, represent an important source of nutrition, especially for populations that are otherwise dependent on a narrow range of staple foods.

1.4.3 China's Contribution in Aquatic Products and Market Share

For nearly half a century, aquatic product output in China has been growing steadily, and currently accounts for about 37% of the total global aquatic product output. In the last decade, 65% of the increase in the global demand for aquatic food can be attributed to China (Nikolik et al., 2018). China's aquatic food consumption has been growing steadily at approximately 5% annually over two decades. The per capita apparent consumption of aquatic food in China reached an all-time high of 49.9 kg in 2016, then slightly decreased to 46.45 kg in 2019, still more than double the global average of 20.5 kg/person (FAO, 2020). Among these, freshwater aquatic products still comprise the majority share of China fish consumption, while seaweed also represents a small share. The growth rate of aquatic product output far exceeds China's relatively stable population growth, indicating that aquatic products are becoming a larger part of the national diet (FAO, 2016). As China's population continues to grow slightly from the current 1.3 billion at an annual rate of 0.5%, its demand for aquatic products should continue to increase (World Bank, 2018).

Chinese aquatic products accounts for 92% of domestic aquatic product consumption (Fig. 1.4). Export of aquatic products is only around 7% of the total production in volume, though China's seafood export is at the high-value end. In 2019, the total export of aquatic products was 4.27 MT in volume and US$20.7 billion in value, representing almost 30% of China's gross export of agricultural products. By comparison, the import value of aquatic products was US$18.7 billion, at 25% increase from the previous year (BOF, 2020). China's demand for seafood and other high-value marine products has grown in recent years, and China is increasingly importing seafood, especially high-value products from foreign sources (MARA, 2018). Despite the fact that fishmeal still comprises a large share of these imports, the share of high-value premium catch fishery products has been on the rise for several consecutive years. Boston lobsters, Dungeness crab, Atlantic cod and many other foreign species are very popular in China seafood markets.

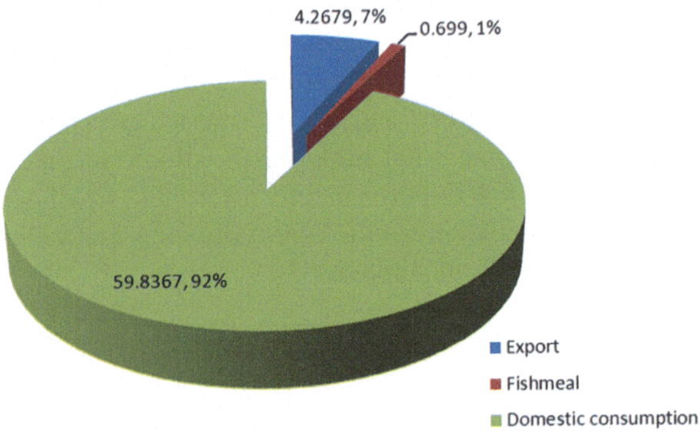

Fig. 1.4 Markets for Aquatic Products in China, 2019 (data in MT and percentage) (*Source* BOF, 2020)

1.4.4 Structure of China's Market for Aquatic Products

In recent years, China's rapid economic growth has brought about steady and rapid growth in residents' income, and food consumption has also begun to transform from meeting basic needs to multi-level, quality needs. There is a significant change in the eating habits of the Chinese people, as "healthy foodstuffs", such as fruits and vegetables, and seafood in particular, have gained increasing shares. In the diet composition of Chinese households, grain (the traditional "main food") has decreased from 235.02 kg/person per year in 1983 to 119.32 kg/person per year in 2012, a decrease of 49.2% (Luo & Xu, 2017). Over the years, China's per capita consumption level of aquatic products has increased significantly, with an increase of 267.62% from 1983 to 2014 (Luo and Xu, 2017). With the decline of grain consumption, eating has shifted focus from mere subsistence to nutrition. Based on the above analysis, and the estimation by FAO (2016) and World Bank (2018), China's demand for aquatic food supply will continue to increase.

Chinese consumers' demand for aquatic products is very diverse, with fish, shrimp and shellfish on offer in Chinese markets alongside high-quality aquatic products from around the world. Seafood consumption patterns in the central, eastern and western regions display significant regional characteristics, due to the influence of natural environment, culture and cooking habits (Zhang & Zhao, 2018). In western China, aquatic products consumption focuses mainly on fish, which is as high as 59%, while the proportion in the central and eastern regions is slightly lower. Shrimp consumption is the largest in eastern regions, accounting for 27%, while in central and western China the ratios are 24 and 21%, respectively. Although China is the largest mollusk and algae producing country, the consumption of these products is generally <10%; the consumption of algae in the east is low, simply because of the availability of a rich variety of aquatic products. With the rapid development of China's distant water catch fishery, and more and more yield being shipped back to China, there are more varieties of aquatic products, such as squid, on the market, and their consumption now accounts for about 5% of aquatic products. Before 2000, China's imports of aquatic products were only about 1–1.5 MT; after China's entry into the World Trade Organization, the scale of imports showed significant growth. In recent years, China has been importing millions of tons of aquatic products for consumption from Russia, the United States, Norway, Canada, Peru, Chile and ASEAN countries every year. Middle and high-end aquatic products, such as Norwegian salmon and Canadian king crab, have greatly enriched the effective supply of Chinese aquatic product market (Sun et al., 2017).

However, unlike some nations with high rates of aquatic food consumption, China's high rate of consumption does not necessarily reflect strong dependence on aquatic food for nutrition or food security. In 2011, finfish comprised only 2% of the diet of China's rural population and 5% of the diet of the urban population, whereas the percentage of land-based sources of protein was nearly three times higher in either population (Liu, 2014). This rural/urban divide suggests that increasing aquatic food consumption reflects growing affluence in China and preference for, rather than dependence on, higher quality food. Growth of China's middle class; improved product handling, storage and transportation infrastructure; and access to new markets are important drivers of changes in Chinese consumer preferences. It is estimated that by 2030, the total demand for aquatic products in China will exceed 70–90 MT (Ren et al., 2012; Tang, 2013; Sui et al., 2017).

Four basic factors affect the consumption of aquatic products in China: level of per capita income, price of aquatic products (undoubtedly the most important determining factor), population numbers and level of urbanization. Compared with that of rural residents, food consumption in China's urban population is more diversified, and food related expenditure much higher. Changes in volume and structure of consumption are relatively stable, with grains and vegetables declining rapidly. Urban residents also consume almost double the amount of aquatic food than rural residents before 2004 (Yue et al., 2017). However, this structure seems to have fundamentally changed in the last 20 years or so. From 2004 to 2014, the ratio of consumption of aquatic products by urban residents no longer increased (Fig. 1.5). Even though China's total food protein consumption has been growing, the growth in ratio of seafood consumption was mainly driven by rural areas (Luo & Xu, 2017). However, considering the still significant difference between rural and urban ratios of aquatic food consumption, and the ever increasing urbanization and rural migration to major cities in China, the driving force of the country's diet structure is a strong indicator for an increase in fisheries production.

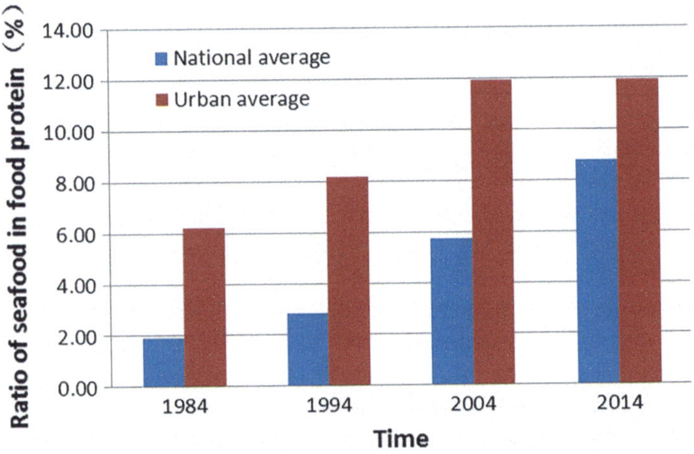

Fig. 1.5 Ratio of aquatic to non-aquatic food in Chinese People's Food Protein (*Data Source* Luo and Xu 2017)

Part I
Status, Trends and Challenges in Management of Living Marine Resources in China

Chapter 2
Mariculture

Hui Liu and Pengyan Zhang

2.1 Marine Aquaculture Production in China

China currently contributes 57.9% of global aquaculture production, and produces considerably more aquatic products through aquaculture than wild capture fisheries in both marine and freshwater systems (FAO, 2020). Freshwater products account for around 60% of total aquaculture production in China, but mariculture production is still substantial and continues to grow. China's mariculture industry has grown markedly for more than six decades (Fig. 2.1), increasing from fewer than 100,000 tonnes in 1950, to 12.64 MT in 2006—surpassing marine capture fisheries—and reached the all-time high of 20.65 MT—double the production of marine capture fisheries in volume—in 2019. During this rapid growth, both marine and freshwater aquaculture have made significant contributions to maintaining domestic aquatic food supply, boosting rural economies and fish farmers' income, improving diets and safeguarding food and nutrition security. During the decade from 2006 to 2016, China's marine aquaculture production increased from 12.64 to 19.63 MT, accounting for 50–56% of China's total seafood output. Then, following stricter law enforcement and the implementation of the national policy of "double control" on fishing power and fishing boats, marine catch fishery yield began to decrease, reaching 10 MT in 2019, representing almost a quarter's decrease from that of 2016. To counterbalance this decreased output, mariculture production increased steadily and reached a new record of 20.65 MT in 2019.

China's mariculture industry is different from that of other countries, in terms of its species composition, style and mode of operations; as well as the vast space

H. Liu (✉) · P. Zhang
Yellow Sea Fisheries Research Institute, Chinese Academy of Fishery Sciences, Qingdao, China
e-mail: liuhui@ysfri.ac.cn

P. Zhang
e-mail: zhangpy@ysfri.ac.cn

H. Liu et al. (eds.), *Conservation and Sustainable Use of Living Marine Resources and Biodiversity*, Global Ocean Governance and Ecological Civilization,
https://doi.org/10.1007/978-981-19-4790-2_2

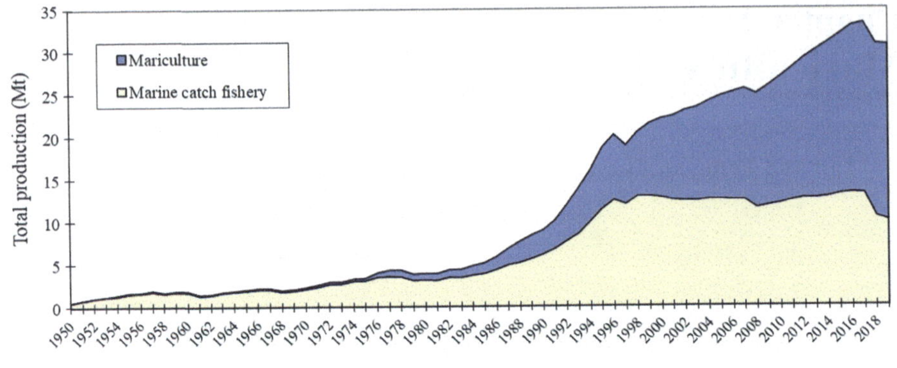

Fig. 2.1 Marine fisheries and aquaculture production in China, 1950–2019 (*Source* BOF, 2020)

and type of marine habitats it occupies. Whereas major aquaculture countries, such as Norway, rely mainly on one or a few dominant species, and a relatively simple mode of operation, China boasts a diversified range of cultured species and scales of operation. Slightly more than 70 mariculture organisms are official registered in the annual statistics (BOF, 2020), including finfish, shellfish, seaweed and sea cucumbers. A considerable proportion of them are grown by photosynthesis or filter feeding on plankton, with no artificial feeding during the culture process. Only a certain percentage of finfish and crustaceans are fed species, and their total production accounts for about 15% of the total mariculture output (Fig. 2.2).

Backed by powerful policy, and fiscal and technological support, China's mariculture has seen robust growth for over half a century and become an important driver of the rapid development of the ocean economy in past decades. Since 1950, China's aquaculture industry has enjoyed development opportunities in inland and coastal areas because the central government has adopted a "cultivation-focused", or "aquaculture priority", development policy, resulting in production doubling every

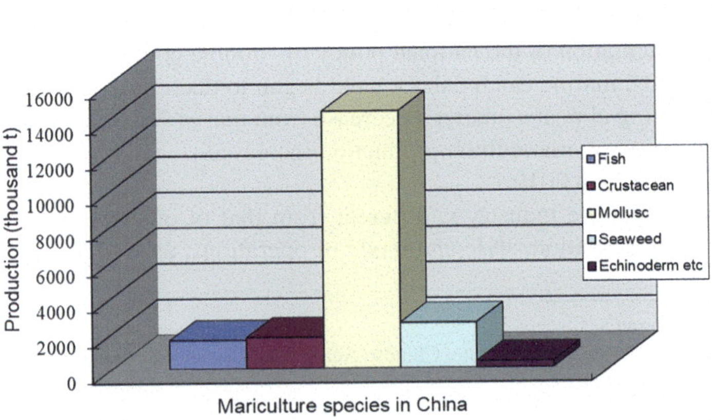

Fig. 2.2 Production of respective mariculture species in China, 2019 (*Source* BOF, 2020)

four to five years. For mariculture, the "five tides", or great leaps forward in the scale of development motivated by the adoption of new species, have greatly accelerated the expansion of production and diversification of products. Backed by favorable policies and government-funded research, these leaps were further enabled by technical breakthroughs in the hatchery and grow-out of kelp, scallops and shrimps; large-scale seedling production of marine fishes; and, more recently, breakthroughs in cultivation methodologies for sea cucumbers and other high-value species. Since 1990, China has made considerable progress in genetic breeding of new varieties and strains of aquaculture organisms; aquatic disease control; culture technology optimization; and harvest mechanization, due to technological improvements made possible by major research project funding from the central and provincial governments. Currently there are thousands of research teams from Chinese universities and research institutes working on a full range of aquaculture-related topics.

Since the 12th Five-Year Plan period (2011–2015), China has transformed its aquaculture policies by highlighting sustainable development. In April 2015, the China State Council issued the Water Pollution Prevention Action Plan, which clearly stipulates that "an ecologically healthy aquaculture will be promoted, with aquaculture prohibition areas drawn for key rivers, lakes and coastal seas; upgrading of aquaculture facilities; strengthening the control over feeds and chemicals; and encouraging deep water aquaculture practices". An upper area limit of 2.2 million ha was also set for mariculture, while "volume reduction & value increase, quality and efficiency improvement, and green development" were set as targets for 2020 by the 13th Five-Year Plan for Fisheries and more recent regulations. In the long run, China's policy on mariculture will be "healthiness and sustainability", with both ecological and food safety implications, and relevant laws and regulations have already been enacted to guarantee implementation.

The environmental impacts of aquaculture (e.g., waste discharge, misapplication of chemicals, and occupation of vast coastal wetlands) have drawn marked public concern in China over the last few years, as awareness of ecosystems and environmental health increases; progress is noticeable in the mass closure of coastal farms and the appearance of extensive coastal restoration projects. The clear division of responsibility in the new regulations, as well as the reorganization of ministries and shift in China's administration in 2017–2018, promises to significantly improve ocean and mariculture governance.

2.2 Ecological Impact of Mariculture

Aquaculture can be a relatively low pollution, carbon neutral (taking into account the mutual offset of carbon absorption and emissions in multi-species polyculture) and environmentally friendly food-producing industry, given all aquaculture activities are carried out scientifically and at a reasonable scale. The ecological impacts of mariculture mainly include habitat encroachment, environmental pollution and displacement of wild living marine resources. The lack of scientific and rational spatial planning for mariculture is the main cause for many of these problems, which are rooted in policy gaps and manifest most significantly in overcapacity of many mariculture waters in China.

2.2.1 *Overcapacity*

The marine aquaculture industry in China is mainly managed through the issuance of a sea use license and aquaculture permit. After obtaining these two documents, the farm owner can engage in aquaculture activities in the sea area where the rights are confirmed. Although the two certificates clearly specify the water space that can be used, there are no restrictions on culture density, cultured species and layout of culture facilities (Liu, 2016). In the case of rising prices of aquatic products, farmers tend to either favor more profitable species or increase the density of seedlings. It is difficult for local authorities to track exactly which species each company is stocking, and take corresponding measures to control the production and waste discharge in specific waters areas; therefore, changes in cultured species or the expansion of the scale of operations are often not subject to proper monitoring, supervision and restriction. Before the 1990s, this tolerant management approach played an important role in promoting the development of China's aquaculture industry. However, in certain coastal regions, the continuous expansion of both the space and scale of aquaculture has led to an unrestricted increase in the biomass and density of aquaculture per unit of water space. This, in turn, has led to increased pollution of the aquaculture systems, a declining environmental quality, more frequent disease outbreaks, and difficulties in guaranteeing the quality of aquatic products. In addition, due to conflicts over the use of aquaculture waters, it is not uncommon for farmers to engage in marculture activities without a license, which in turn causes overcapacity. Given the loopholes in current laws and regulations, and especially weak law enforcement, China's fishery management authorities need powerful spatial planning tools to promote strict management of mariculture.

2.2.2 *Encroaching Upon Wetlands and Habitats*

Though mariculture industry is an important and reliable contributor for China's food security and coastal economic development, it requires a large amount of physical space. Coupled with overcapacity in aquaculture, these can be seen as two sides of the same problem. Mariculture occupies about one-third of the coastal wetland area and 10% of shallow sea space in China (Liu & Su, 2017). In addition, China has also witnessed large-scale sea reclamation for mariculture (Wang et al., 2014), including cofferdam and earthen ponds, which has altered a large expanse of coastal landforms and degraded coastal wetland ecosystems. Approximately 240,000 ha of shrimp ponds have been built in the coastal areas of southeastern China over the past 40 years, largely by destruction of mangroves and seagrass beds. According to the recent nationwide marine inspection (SOA, 2018), reclamation and mariculture development in many provinces has been in violation of laws and regulations, but without adequate restriction by the local authorities. For example, in Hebei Province, the total area of shallow sea and reclaimed wetlands used for mariculture is around 18,424 ha, only 27% of which is licensed. In Jiangsu Province, unlicensed mariculture operations take place in 137 sites and cover around 12,000 ha, with 9,954 ha in conflict with Marine Protected Area (MPA) buffer zones and Ecological Redline protected areas. Between 1989 and 2000, China lost 12,924 ha of mangrove forests, more than 97% of which was due to the construction of shrimp ponds. Open violation of law and regulations in mariculture practice is not rare, clearly reflecting weakness in law enforcement.

2.2.3 *Waste Discharge and Pollution*

Overcapacity in mariculture operations not only displaces many natural habitats, but also degrades remaining habitats through pollution. According to the 2015 report by the National Fisheries Environmental Monitoring Network (MOA & MEP, 2015), pollution of some coastal mariculture water bodies in the four China Seas remains severe, while the exceed-limit ratio for inorganic nitrogen, labile phosphate, chemical oxygen demand (COD) and petroleum hydrocarbon in all samples was 78.1, 54.9, 13.1 and 16.2%, respectively, and mariculture was identified as one major source of all these pollutants (Jia et al., 2017). Influenced by riverine discharge and aquaculture, antibiotic pollution had been increasing in some coastal waters, including the Beibu Gulf (Zheng et al., 2012), where erythromycin was detected in 100% of samples at concentrations ranging from 1.10 to 50.9 ng/L, and sulfamethoxazole pollution was detected in 97% of samples at concentrations up to 10.4 ng/L. In recent years, China has initiated actions for general pollution interception/emission reduction and aquaculture waste water treatment, and MARA has banned the application of antibiotics in aquaculture in recent years (MARA, 2019), the quality of fisheries waters has already been significantly improved (MARA & MEE, 2018).

Most of the mariculture output in China is from extractive species, such as seaweed and mollusks, yet fed species, including finfish and crustaceans, are also cultured in large quantities (total volume at 3.35 MT in 2019) (Fig. 2.2). For fish culture, only about 27–28% of the nitrogen given as feed was redeemed as fish, and more than 70% of nitrogen is released into the environment (Hall et al., 1992). These species clearly contribute to marine pollution, especially when they are cultured in both large scale and high density. Nantong City is among the major whiteleg shrimp *Litopenaeus vannamei* production sites in China; the scale of shrimp culture has expanded from 6700 ha in 2013 to 12,700 ha in 2017, almost doubling production within four years. The rapid spread of small sheds for shrimp culture across mudflats and farmland has led to a series of environmental problems, such as soil salinization and over-extracting and polluting of shallow groundwater resources. Similar issues are also evident in major aquaculture areas in Hebei, Shandong and other provinces, and the cleaning and treatment of mariculture-discharged pollutants has become an important challenge for local governments. Control of aquaculture discharge is one major responsibility of fisheries authorities at all levels, according to the Fisheries Law of China. However, for a long period of time, this task has been largely neglected or given way to enhancing production; lack of monitoring expertise and waste discharge standards may partially share the blame. Still, a recent study on effluent of nitrogen and phosphorus by net cage culture of marine finfish shows that the nitrogen and phosphorus pollutants generated by aquaculture (1000 kg production) was 72 and 17.3 kg if formulated feed is used, which was significantly lower than that generated from trash fish as feed (142 and 26 kg, respectively). Compared with inputs from atmospheric subsidence, rivers, submarine groundwater and sediment interface exchange, nitrogen and phosphorus input from cage culture was found to be relatively minimal, accounting for approximately 7 and 3% of the total input, respectively (Qi et al., 2019). Though it is incorrect to attribute the problem of coastal eutrophication solely to aquaculture, the impact of aquaculture activities on the increased nutrients and organic pollutants in the local waters should not be ignored.

2.2.4 Pressure on Capture Fisheries

An additional concern about the adverse environmental impacts of mariculture is the high volume of feed needed to produce certain fed species (such as fish and crustaceans), often derived from wild capture fisheries harvesting ecologically important forage fish. Some species of farmed fish still use fresh trash fish as the main feed, because formulated feed with good palatability has not yet been developed for them. For example, China produced 225.5 thousand tons of large yellow croaker in 2019. This species is fed almost nothing but trash fish, and its feed conversion ratio (FCR) is 7:1. It is thus estimated that, in 2019, nearly 1.5 MT of China's trash fish catch were used to feed only one species of cultured fish. China's aquatic feed industry has developed rapidly for more than 30 years; its production increased nearly 30-fold, from less than one MT in 1991 to more than 22 MT in 2018, representing over 50% of total global production (Mai, 2020). Accompanying this growth has been the development of the world's largest aquatic feed manufacturing enterprises, as the processing techniques and quality of some aquatic feeds produced in China have improved over time. For instance, the feed coefficient of prawn has dropped to 1.0–1.2, nearly meeting the international standard for efficient production (Liu et al., 2017). However, in some cases the feed coefficient, or the ratio of feed inputs to output of mariculture products, is still very high. Excessive use of feeds can exacerbate pollution and habitat destruction caused by other aspects of mariculture operations. Scientific and technological progress in genetic breeding, facility upgrading and automation is very important for improving feed efficiency and the comprehensive performance of aquaculture industry.

2.3 Social and Economic Factors

2.3.1 Socioeconomic Development Pushes Up the Demand for Aquatic Products

By 2030, China's population is expected to reach 1.5 billion, and the total demand for aquatic products will probably reach a record of 70 MT (Tang, 2013), if its apparent consumption is calculated on the basis of 46.5 kg (per capita, statistics in 2018). However, as the birth rate in China further declines and societal aging intensifies, China's total population may fall back to between 1.2 and 1.3 billion by 2050, and China's demand for aquatic products will remain close to current levels. However, from a global perspective, population growth (an annual increase of 1.6%) and the corresponding increase in demand for aquatic products (kept at the annual increase of 3.1% between 1961 and 2017) (FAO, 2020) are likely to remain until the middle of this century, and the world is likely to have to sustain a population of more than nine billion, whereas the total demand for aquatic products will also exceed 300 MT. Therefore, maintaining sustained and steady growth in China's production of aquatic products is not only of vital importance to China, but also of particular importance to helping secure world food supply.

Since the late 1980s, mariculture has become an active player in China's marine economy, partly in response to the continuous decline in marine capture fisheries yield over the prior decades. Considering its continuously growing supply and high value of seafood, mariculture is undoubtedly a major contributor to China's food security and economic growth. At present, in the face of new development needs and the pressure of ecological environmental protection, while freshwater aquaculture is strictly limited by space and productivity increase in the short term, mariculture must continue to expand into new spaces and new industrial forms, and improve its intensification and production efficiency.

To solve the problem of limited space in tidal flats and inshore aquaculture, China has continuously explored deep-sea aquaculture engineering and intelligent technology; more than 10 new types of deep-sea aquaculture facility have been successively established, at a total volume of over 300,000 m^3 (Fig. 2.3). Relevant deep-sea aquaculture technology for premium marine fish species, such as salmon, yellowtail, rock bream, yellow fin tuna, large yellow croaker, grouper and cobia, has also been developed (CAST & CFS, 2020).

1. 3000 ton mariculture factory vessel (at cold water mass off Rizhao);
2. 50,000 ton submersible fish farm "Deep Blue No. 1" (at cold water mass off Rizhao);
3. 9000 ton large steel structure cage (at sea area outside Port Tieshan, Guangxi);
4. 30,000 ton "Dehai No. 1" intelligent fish farm (at Wanshan sea area, Zhuhai);
5. 160,000 ton large-scale pipe-and-stake aquaculture encampment (in Laizhou Bay);

Fig. 2.3 Deep-sea mariculture facilities developed and currently in use (trial) in China (*Source* Guan CT (CAST & CFS, 2020))

6. 30,000 ton offshore island reef foundation net cage (at Meiji Reef, South China Sea);
7. 60,000 ton "Long Whale No. 1" bottom-mounted smart cage (at Changdao sea area, Yantai);
8. 13,000 ton "Zhenyu No. 1" self-turning cage (at Lianjiang sea area, Fuzhou).

Recreational fisheries are a relatively new industrial form in China, closely related to tourism and aquaculture. Since the implementation of recreational fisheries monitoring statistics in 2003, the value and proportion of recreational fisheries in total fisheries economy have shown an overall upward trend in China's inland and coastal provinces, with an average annual growth rate of 19.56%. In 2019, the total output value of recreational fisheries reached 96.37 billion CNY, accounting for about 3.6% in value of China's fisheries economy (BOF, 2020; NCSFTP, 2020). With the expansion of China's sea ranching projects, the proportion of marine-related recreational fisheries is expected to further increase.

2.3.2 *Considerable Risks Imposed by Disastrous Events*

As an industry that close to and relies on the sea, mariculture faces a number of major challenges, including labor shortages, rising feed and energy costs, and frequent occurrences of natural disasters and disease outbreaks. The revenue of fisheries economy generated by China's 11 coastal provinces and cities contributes 71% of the country's total fisheries output value. It is thus estimated that the annual economic loss of fisheries in coastal provinces and cities caused by natural disasters was also approximately 70% of the national total (Table 2.1). These figures may well represent the loss of China's mariculture industry, although freshwater aquaculture may contribute significantly in some of these provinces. In 2018, an economic loss of more than nine billion CNY was reported for the 11 coastal provinces and cities, caused by typhoon and flooding, disease, drought, pollution and other natural disasters.

Fisheries economic loss in the coastal provinces due to environmental and water pollution ranked first among all causes, and accounted for 73.4% of the national total fisheries loss due to pollution (slightly higher than the percentage of fisheries economy of these regions). This may indicate the situation of pollution in China's mariculture waters, as disastrous pollution accidents still occur frequently, resulting in mass mortality of cultured animals and huge economic losses. For example, on November 4, 2018, in the Quangang District of Quanzhou City, Fujian Province, a ship owned by Donggang Petrochemical Industry Co., Ltd. had a leak in the hose connecting it to the dock, resulting in 6.97 tonnes of carbon-nine spill into the surrounding sea. This accident directly resulted in millions of CNY loss for nearby fish farms (Phoenix News Media, 2018).

On the other hand, coastal water eutrophication is apparently shown in the increased occurrence of algal blooms and green tides over the last decade. In June and July 2019, a huge biomass of green tide species *Ulva prolifera* landed along the

Table 2.1 Economic loss of fisheries in Coastal provinces and cities caused by disasters of (measured in 10,000 CNY)

Provinces & cities 地区	Subtotal 小计	Typhoon & flooding 台风、洪涝	Disease 病害	Draught 干旱	Pollution 污染	Others 其他
Tianjin 天津	1035		1035			
Hebei 河北	126,336.1	7757	3154.1		1380	114,045
Liaoning 辽宁	235,568.6	238	9048	102,503		123,779.6
Shanghai 上海	1889.58	331	1457		101.58	
Jiangsu 江苏	145,306	40,454	52,937	3476	33,049	15,390
Zhejiang 浙江	48,976	24,126	23,015	431	1292	112
Fujian 福建	56,438	34,954	9284	1378	10,090	732
Shandong 山东	143,737	56,471	18,434	6634	13,300	48,898
Guangdong 广东	173,285.35	127,830.16	25,389.27	1744.8	953.76	17,367.36
Guangxi 广西	10,795.26	2952.02	5739.74	108.6	133.1	1861.8
Hainan 海南	2047.8	1040	964.8		43	
Total coastal regions 沿海合计	945,414.69	296,153.18	150,457.91	116,275.4	60,342.44	322,185.76
National total 全国总计	**1,363,365.93**	**475,108.34**	**261,259.43**	**193,329.9**	**82,211.46**	**351,456.78**
Percentage of coastal regions 沿海占比	69.34%	62.33%	57.59%	60.14%	73.40%	91.67%

Source BOF (2019)

Fig. 2.4 Large scale green tide of *Ulva Prolifera* occurred in coastal waters of Qingdao City (photo by Jiang XP on July 20, 2019). Research vessel on upper-left is 73 m in length

coasts of Shandong Province, causing pollution of mariculture waters and bathing beaches. Thousands of fishing boats, trucks and bulldozers, as well as army soldiers, sanitation workers and volunteers, were called upon to spend days collecting and removing this pollutant (Fig. 2.4). Besides the trouble of *Ulva prolifera* for 13 consecutive years, reoccurrences of outbursts of floating seaweed *Sargassum horneri* have also happened in recent years. In May 2019, a large biomass of floating *S. horneri* was observed in the northern shoal of Jiangsu Province, where it piled upon and destroyed the laver rafts (Science News, 2019), before moving northward in June to further pollute the offshore water. Indeed, human-induced and natural disasters alike pose a severe threat to the healthy development of mariculture industry.

2.3.3 Significant Social and Economic Responsibilities

In 2019, China's mariculture industry generated 20.65 MT of product and a total value of 357.5 billion CNY, provided nearly 2 million jobs, and supported an entire chain of seafood processing, transportation, catering and marketing, and other service sectors (BOF, 2020). At present, the total output value of marine fisheries accounts for about 14% of the added value of China's marine economy. In short, mariculture bears important responsibilities in terms of food security, economic growth and fishermen's livelihoods. China must work hard to ensure the sustainable and healthy development of the mariculture industry.

In addition to economic losses caused by natural hazards and disasters, a significant increase in the prices of feed, energy and labor in recent years has eroded what was once reliable profitability in the early days of mariculture development. Wide price fluctuations of a number of mariculture products, and declining consumer confidence shaken by occasional product safety incidents (Chen & Chen, 2006), have further exacerbated these impacts. At the beginning of 2020, the outbreak of COVID-19 dealt a heavy blow to the aquaculture industry: prices, sales and trade volume of aquatic products have all dropped significantly; the sales price of China's marine finfish, import volume of shrimp and salmon, and global import and export of aquatic products all declined. Generally speaking, the aquatic product industry is more sensitive to the macroeconomic situation; improving resilience against external risks, as well as improving stability in profits, employment, product quality and environmental friendliness, is possibly the most important challenge facing the industry today.

In China, large-scale mariculture has gradually become part of the coastal ocean ecosystem; thus, the integrated management of aquaculture, capture fisheries and habitat protection is becoming more and more important. In order to reestablish the coastal habitats and increase fisheries stocks, as well as boost the marine fisheries economy, China has invested heavily in sea ranching (defined in China basically as artificial reef construction and stock-enhancement efforts) at a large scale over the last few years, identifying it as a new style of enhancement-based fisheries. However, due to the lack of systematic research and knowledge, and lack of proper supervision and governance, sea ranching may also fail ecologically and economically. Similar to aquaculture, it is highly important to conduct scientific and technological research, and strengthening environmental monitoring and governance on sea ranches, in order to protect the industry from hazards and disastrous events. By the end of 2020, China has built six batches of 136 national sea ranch demonstration sites, with new measures such as scientific site selection, risk assessment, ecological monitoring and intelligent management, increasingly adopted. It is believed that under the general framework of comprehensive management of marine biological resources, China's marine aquaculture industry will develop steadily in the future.

2.4 Technical Challenges

2.4.1 Industrial Technology

During the long history of mariculture development in China, there have been numerous technological breakthroughs and innovations; for example, in commercial scale seedling production, growout facilities and operational tools, to name a few. According to CAST and CFS (2020), in recent years, the preservation and identification of marine genetic resources in China, and the cultivation of new varieties, have achieved remarkable results. The genomes of a number of marine species have been deciphered, and the molecular mechanisms of their important economic traits analyzed. Stem cell transplantation technology, genome editing technology and genome selective breeding techniques have seen further improvement and application in marine organisms. Offshore mariculture net cages and land-based mariculture facilities have been further upgraded. As a result of the rapid development of deep-sea aquaculture facilities and equipment, China is one of few countries around the world to own large farming facilities, such as the "Deep Blue No. 1" deep sea net cage for culture of finfish. A large number of studies have been carried out on protein nutrition and its substitution, lipid nutrition and substitution, vitamin and mineral nutrition, feed additives for mariculture species, efficient environmentally friendly feed development, and a series of important research results have also been achieved for mariculture artificial feed formulation. Significant progress has been made in the pathogens and epidemiological studies of new diseases for mariculture species; breakthroughs were made in the development of multiple quantitative and rapid detection techniques for various aquatic pathogens, and research on aquatic pathogen infection and host immune mechanisms, aquatic vaccine development and pharmacology screening studies has also achieved positive results.

From a general perspective, facing the intensified evolution of marine ecosystems, the most urgent scientific and technological issues related to China's living marine resources are undoubtedly multifaceted. It is important to understand the current characteristics, evolutionary trends and driving mechanisms of land and ocean ecosystem interactions under the influence of human activities; coordinate and link the technical methods of coastal land and ocean environmental monitoring and ecological restoration; and scientifically and rationally use and restore living marine resources, in order to establish a new mode of coastal ecosystem protection and sustainable utilization.

In consideration of China's mariculture technology, there are challenges to be faced in all aquaculture modes, including pond, industrial, inshore and offshore mariculture.

(1) **Pond mariculture**
Inadequate research on the mechanism of formation and variation of pond ecology and its key influential factors, has restricted timely control of pond aquaculture. Consequently, pond farming now mainly relies on manpower, level of mechanization is relatively low, and there is a shortage of mechanical equipment, aside from oxygenation and feeding machines.

(2) **Industrial mariculture**
Industrialized recirculating aquaculture systems (RAS) now have been widely used all over China, and the performance of these systems (e.g., productivity, energy comsumption, and water reuse rate) is rather advanced at the international level. However, certain gaps still exist in system integration, performance of machinery, and standardization.

(3) **Inshore mariculture**
The lack of standardized design of inshore aquaculture facilities in China has greatly impeded its mechanization upgrading. Despite a few processes being carried out by machinery, manual operation is still widely used, labeling these mariculture modes with high labor intensity and low production efficiency.

(4) **Offshore mariculture**
The deep sea is characterized by special weather and hydrodynamic conditions. Long distances from the coast, strong oceanic currents, storm surges and other disastrous weather pose more severe challenges to the adaptability and reliability of structural layout, as well as to the exchange performance of caged-in water bodies of deep-sea aquaculture facilities. Further basic research and solutions application is needed in this area.

2.4.2 *Governance Measures*

Since the 1980s, China has been establishing a legal system for governance of marine fisheries and aquaculture, with the *Fisheries Law* as the main body, and the *Sea Area Use Law* and the *Marine Environmental Protection Law* as the main supportive laws, together forming a framework of five national laws and four administrative regulations, as well as several rules and regulatory documents. Still, given these relatively complete policies and regulations, China has yet to improve on fisheries resource management and conservation approach and measures. Although China's aquaculture management policies and regulations are relatively complete, management measures and technical means still need to be improved; quite often, there is insufficient technological research and development to support management of such a diverse and gigantic industry. Long-term, continuous aquaculture environment monitoring and production data collection are the most important tools for mariculture governance. However, neither process is typically comprehensive nor timely, which prevents management authorities from detecting problems and taking proactive measures. Information system construction for fisheries governance lags behind modern information technology, and mariculture licensing is not strictly attached to all fish farms. Local fisheries authorities are urgently in need of a farm-based, routine stock and disease reporting system to facilitate rapid response and speed up management actions. As a result, the local Ocean and Fisheries Bureaus (now renamed as Natural Resource Bureau or Ocean Development Bureau in some provinces and cities) usually do not know exactly which species are cultured and how much is produced at respective fish farms, nor do they know the exact quantity of feed and chemicals being used by all the farms. There has not been enough monitoring of mariculture's impact on the environment, or conversely, the environmental factors affecting mariculture—including the amount of discharge by respective farms, or which farms in particular are more susceptible to pollution—partly due to the difficulty in data sharing between government authorities and research institutions. This lack of data, combined with the lack of data-supported decision making, creates a hindrance for effective mariculture governance.

Environmental and fisheries policy in China has evolved significantly during the past decade to confront many of the environmental impacts of mariculture. Since the 18th CPC (Communist Party of China) National Congress in 2012, China has elevated its commitment to ecological civilization. As a result, China's environmental quality has improved, especially as a result of the 2015 National Environmental Protection Law, which saw environmental supervision, ecological auditing, environmental law enforcement and punishment for violation all strengthened or increased in due course. Yet, after decades of unconstrained investment and unrestricted growth, mariculture undoubtedly became confronted with unprecedented pressure imposed by these policy changes. The mariculture industry now faces a ban on sewage and sludge discharge and the use of coal-fired boilers; some fish farms are required to retreat once a conflict is identified with an MPA or the Ecological Redline system, tourism, or certain other sea uses. Mariculture enterprises must decide whether to disband or reform their operations to comply. China's commitment to improving environmental quality and reducing carbon emissions will affect all economic sectors, including mariculture; as this not only poses an imminent burden on the industry, but also provides a new opportunity for mariculture to transform. Swift action should be taken to improve technology to enable less carbon-intensive pollution treatment and green development.

Furthermore, according to the 13th Five-Year Plan for China's Fisheries Industry and subsequent national policies, "withdrawing aquaculture and returning to natural tidal flats" is a general policy for mariculture in the near future. Balancing spatial restrictions with continuous provision of high-quality seafood has become the greatest challenge for fisheries authorities, researchers and fish farmers, and can only be solved via technical breakthroughs in all sectors of China's mariculture. China is close to launching a Territorial Spatial Planning initiative covering all land and coastal ocean areas, which will also be important for aquaculture. Therefore, one of the most important and imminent tasks is to rationally draw the layout of mariculture ponds, industrialized facilities, offshore and deep-sea aquaculture farms, by taking into account the ecological and economic requirements. Fisheries administrative authorities should take advantage of this initiative and formulate implementation plans in advance, so as to clarify the relationship between aquaculture and local economic, social and ecological development.

2.5 Summary and Conclusions

China's mariculture industry is extensive in scale; undoubtedly it makes great contribution to the social economy of not only the coastal areas but also the whole nation. However, the development of mariculture also faces a series of challenges, including the internal issues such as costs, labor, and market price, as well as externalities such as carrying capacity, raw materials (feed and input) and waste discharge. Some of these are problems that naturally arise in the process of industrial development, which need to be continuously dealt with and resolved through technological research and innovation, transformation and upgrading; some are management problems that need to be resolved by improving policies and regulations and strengthening enforcement and governance measures. At present, China is vigorously promoting the green development of aquaculture. In view of China's status as a major country in the global and regional aquaculture industry, China's development of ecologically healthy mariculture and high-quality, green and ecological aquatic products will provide an important model for global industrial development and seafood supply.

Chapter 3
Fisheries

Jacob P. Kritzer, Fang Sun, Daniel Willard, John Mimikakis, and John Virdin

Overfishing is not only the most significant problem facing many fisheries worldwide, it is perhaps the most significant issue facing marine ecosystems overall. This is not to detract from other severe issues that exist in some locales; however, because fishing fleets are vast and widespread, and cause direct mortality to harvested organisms—often with incidental impacts on habitat—overfishing remains a major threat to ocean health at the global scale, rivaled only by climate change (Halpern et al., 2008). Indeed, recent estimates suggest that ending overfishing can promote recovery of marine wildlife, including mammals, birds and turtles, illustrating the severe ecosystem-level effects of overfishing (Burgess et al., 2018).

China is no exception to the severe consequences and drivers of overfishing. Marine ecoregions in China were once world famous for rich fishery resources and high-quality seafood products. However, overfishing, combined with other sources of deterioration of the marine environment over the past decades, has resulted in increasing occurrences of the "rare fish in the sea" phenomenon, whereby species that were once major components of China's fishery yield have become infrequent in both the catch and the ecosystem. For example, fishing grounds in the Zhoushan

J. P. Kritzer (✉) · J. Virdin
Environmental Defense Fund, Boston, USA
e-mail: jake@neracoos.org

J. Virdin
e-mail: john.virdin@duke.edu

F. Sun
Environmental Defense Fund, Beijing, China
e-mail: sunfang@cet.net.cn

D. Willard
Environmental Defense Fund, Austin, USA

J. Mimikakis
Environmental Defense Fund, Washington, DC, USA

H. Liu et al. (eds.), *Conservation and Sustainable Use of Living Marine Resources and Biodiversity*, Global Ocean Governance and Ecological Civilization,
https://doi.org/10.1007/978-981-19-4790-2_3

Archipelago, covering an area of 220,000 km^2, were once so productive that the region came to be known as the home of fishing in China and the granary of the East China Sea. However, catch of large yellow croaker in the region that reached 170,000 tons in 1957 dropped to only 400 tons in 2015, a decline of more than 99%.

Declining trends in other Chinese fisheries have been similar, especially for the highest value species, although losses have been offset by increased catch of more abundant and low-value species (Szuwalski et al., 2017). New opportunities are also emerging in recreational fisheries and distant water fishing. However, if recreational fisheries and the distant water fleet are not managed sustainably, these transitions might simply mean exporting overfishing problems in commercial fleets to other emerging economic sectors.

The causes of overfishing are diverse and vary from fishery to fishery, though two of the most pervasive drivers include economic incentives that are not aligned with necessary environmental outcomes, and fishers being disconnected from decision-making processes, and therefore less likely to accept and comply with regulations. In many cases, a similar combination of factors underlies fishery declines, including unconstrained development of overcapitalized fleets, technical challenges with stock assessment and management, and socioeconomic impacts associated with management reform and resource recovery.

3.1 Marine Fisheries Production in China

By several measures, China is the world's largest fishing nation. It is by far the world's largest producer in capture fisheries, contributing 15% of world marine capture catch in 2018 and 16% of world inland freshwater catch (FAO, 2020). In 2018, China's fishing fleets harvested 13 MT of marine catch, nearly twice that of the next largest fishing nation, Peru. The fleet is distributed among 11 coastal provinces bordering four major seas (Bohai Sea, Yellow Sea, East China Sea and South China Sea), including over 20,000 marine species and 1000 species exploited commercially (Liang et al., 2018). The East China Sea is the most productive of China's marine waters, generating 40% of its total marine catch (Szuwalski et al., 2017). Fisheries catch in China is multispecies, due to the high diversity of marine species in China's ecosystems; predominant use of trawl, stow net, gillnet and other nonselective gears; general lack of harvest limits (Shen & Heino, 2014); and weak law enforcement. Over time, catch composition has shifted from large volumes of high-value species for consumption, to multiple smaller low-value species, with an increasing share (35% of total marine catch) used for animal and aquaculture feed (Zhang et al., 2019).

Fisheries are an important economic driver in China and are connected with myriad seafood markets around the world. China has the world's largest fishing fleet, with over one million workers on 200,000 fishing vessels (Cao et al., 2017) and nearly 15 million workers in the industry (Zhang & Wu, 2017). Catch within China's exclusive economic zone (EEZ) has increased by nearly an order of magnitude since 1950

(Srinivasan et al., 2012), while a distant water fleet targeting high-value catch has developed more recently and now accounts for 18% of China's marine catch (FAO, 2020). This production supports an aquatic products processing industry with 10,000 firms and 22 MT annual output (BOF, 2018). China is also the world's largest exporter of fish and fish products, with exports of US$20 billion in 2016 and 14% market share. China is the world's third largest importer of fish and fish products, with imports of US$8.7 billion in 2016 (FAO, 2018). A significant amount of fisheries trade volume in China is fish that are imported, processed and re-exported to other countries; however, increasing imports over the years signal rising incomes and changes in Chinese consumer preferences toward species that are not locally produced. By the year 2030, China is expected to consume 38% of the world's food fish production (Fabinyi & Liu, 2014). Altogether, given China's position in global seafood supply chains, its world-leading volume of capture fisheries yield, and the size of its fisheries labor force, management of wild fisheries is a critical policy issue for social, economic and environmental reasons.

3.2 Brief History of Marine Fisheries in China

China's emergence as a global fishing power is a relatively recent phenomenon, yet the industry and total marine catch expanded steadily for decades after the founding of the People's Republic of China in 1949. Shen and Heino (2014) outline four major stages from the mid-twentieth century to the present. The first stage, in the 1950s, was characterized by steady economic and technological development, transforming fisheries from a previously underdeveloped and largely artisanal nature to a more significant industrial enterprise. Most catch during this stage was nearshore benthic species with large body size and high commercial value, such as hairtail *Trichiurus lepturus* and large yellow croaker and small yellow croaker *Larimichthys poyactis.* The second stage, in the 1960s, brought full utilization of the majority of China's marine fishery resources. Fishing shifted toward smaller and lower value species, such as anchovy, as catches of traditional species began to decline. The third stage, spanning the 1970 and 1980s, involved the subsequent decline of many stocks, as fishing effort continued to expand unchecked. Total marine catch was estimated around three MT in the mid-1970s, increasing by only 2% per year through the mid-1980s, at levels of one-half of estimated maximum sustainable yield and far below China's seafood self-sufficiency targets (Cao et al., 2017). In 1978, the national government introduced the Reform and Open Door Policy, a program of economic reforms designed to phase in private enterprise and open the economy to foreign trade and investment. Accordingly, China's fisheries and coastal economies quickly developed for exploitation and growth, market orientation and outward expansion. By the dawn of the fourth stage, beginning in the mid-1980s and continuing to the present, the overexploitation and underperformance of China's marine fisheries was sufficiently evident; this ushered in an era of fishery management reforms initially intended to increase production, and later to promote restoration and conservation.

Recognizing that growth in total marine catch had been somewhat stagnant in certain period of time, in 1985 the national government announced a policy directive that, among other measures, promoted the privatization of fishing vessels and expanded fisheries production within China's EEZ. This encouraged improvements in fishing technology and, with growing recognition that fisheries can be a significant economic driver, effort and catch then expanded rapidly. Over the next decade, marine catch increased by 11.8% per year, total catch volume approached 15 MT, and China became the world's largest fishing nation (Cao et al., 2017).

Throughout this time, and despite declining fish populations, catch volume continued to increase and has remained stable through the present. This trend was sustained for several reasons: first, as has been observed in the development of commercial fisheries around the world, food and market demand for seafood generates high levels of fishing effort and rapid technological advance, allowing fleets to pursue diminishing fish stocks with increasing technological power, leading to intense pressure on fish stocks, sometimes to the point of collapse (Clark, 2006). Second, in Chinese waters fleets serially exploited and found markets for younger, smaller and lower-value species, as those previously utilized were depleted (Zhang et al., 2019). Third, it is possible that populations of lower trophic-level species became more abundant after Chinese fleets depleted larger predators, causing sustained growth and increasing catch of prey species (Szuwalski et al., 2017). Finally, beginning in 1985, China started to develop its distant water fleet, which can offset declines in domestic production through vessels fishing on the high seas or in other nations' jurisdictions.

Since 1985, an increasing share of total marine catch comes from offshore and distant-water fleets (Zhang & Wu, 2017). Mallory (2013) argues that the growth in distant water fishing, together with increasing aquaculture production, has been primarily driven by policy goals of job creation and economic growth, rather than by food needs. Initially, the fleet was entirely state owned, but most fishing businesses became privately owned during privatization reforms of state-owned enterprises in the 1990s. Today, China has the largest distant water fishing fleet in the world, with approximately 2700 vessels, based on official reports (BOF, 2020), though some studies argue the fleet is significantly larger and may vary between 3400 and 17,000 vessels (Pauly et al., 2014; Gutierrez et al., 2020), including nearly 1000 vessels registered or "reflagged" in other countries. The fleet's reach includes high seas around the world and the EEZs of 40 partner nations, particularly in West Africa and Southeast Asia, with total estimated marine catch of over 2 MT, or 18% of China's total marine catch (FAO, 2018; BOF, 2015–2020); approximately half of its catch is exported to developed countries. As China's fishing industry turned outward, the fleet's expansion has been facilitated by various policies, beginning in the 1980s, to constrain domestic waters fishing effort and subsidize offshore and distant water fishing (Fig. 3.1).

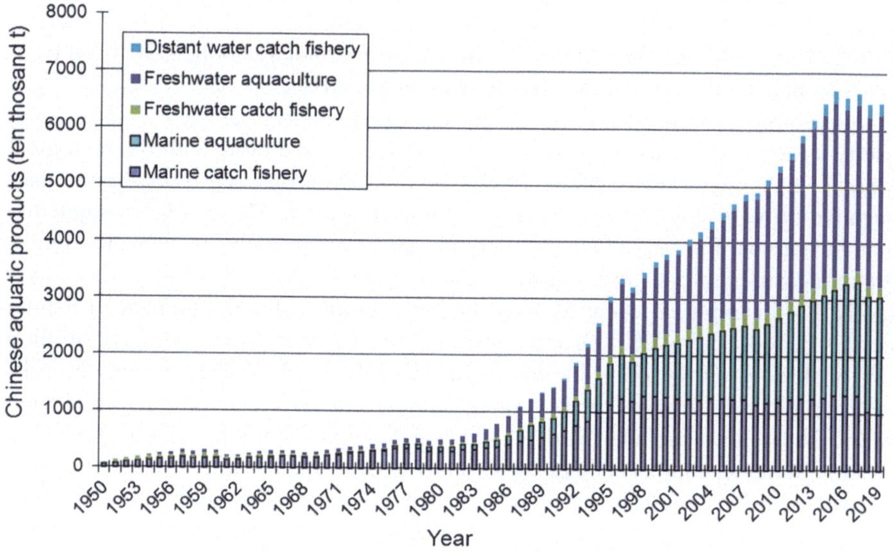

Fig. 3.1 Sources of China's Total Aquatic Production 1950–2019 (*Source* BOF, 1951–2020)

3.3 Regulatory Environment

With the expanded capacity of Chinese fishing fleets in domestic waters generating catch beyond natural limits on fish productivity, government policies to constrain fishing effort became a stronger priority beginning in the 1980s. An early management measure was a fishing licensing system adopted in 1979; while in principle this was premised on the status of fishery resources, in practice it had little impact, but provided a basis for subsequent regulations (Huang & He, 2019). The nation's first comprehensive Fisheries Law was enacted in 1986. This established certain rules for fishing, codifying the licensing system and extending it to all fishing vessels, and establishing authorities for specifying national fishing quotas with respect to available resources.

Administratively, fisheries management is the jurisdiction of the Bureau of Fisheries (BOF) under the Ministry of Agriculture and Rural Affairs (MARA), formerly the Ministry of Agriculture (MOA) prior to government restructuring in 2018, although other agencies and institutes have influence on some aspects of environmental and natural resource management policy. Fisheries policies are guided by Five-Year Plans which have been promulgated sequentially since the early 1950s. Historically, the Five-Year Plans and other policies emphasized production and use of technology to solve problems (Fabinyi & Liu, 2014). Because the country's success in fisheries management was evaluated in terms of annual production, national and local governments prioritized industry development to maximize fishery output and employment opportunities for local communities (Zhang & Wu, 2017). More recently, recognizing the extent of overfishing and habitat degradation in China's waters, priority goals in Five-Year Plans and other policies have emphasized resource conservation and environmental protection, an important symbolic shift from prior goals emphasizing production capacity (Cao et al., 2017).

Accordingly, recent policy changes have aimed to address China's severe overcapacity issues in order to bring nationwide fishing power in line with the biological limits of fishery resources. Since 1999, China has followed zero-growth, and later negative-growth, strategies for fishing in domestic waters. These policies stipulated that the national Bureau of Fisheries and provincial governments were to take appropriate measures to constrain marine catch to 1998 levels, and subsequently lower levels. For example, a recent MARA directive requires that the number of fishing vessels be reduced by 20,000, and cumulative fleet power decreased by 1.5 million kW, while constraining domestic catch to no more than 10 MT. The government seeks to accomplish these goals through production quotas and regulating use of fishing gears. These have included various input and output controls, such as licensing, specific limits on numbers and horsepower of vessels (the so-called "dual controls"), vessel buybacks, closed areas, fishing net mesh size limits, fish minimum size limits, fishermen relocation and stock and habitat enhancement projects. Shen and Heino (2014) conclude that these measures—in particular, mesh sizes and dual controls—generally have had limited impact on fishing effort. Protected areas have been very small and monitoring has been deficient, and studies evaluating their effectiveness are limited. While there have been large-scale efforts to culture and release juvenile fish to enhance fisheries catch—which is now in practice for 100 species of finfish, shellfish and crustaceans—again, monitoring and studies of effectiveness have been limited (Cao et al., 2017). Recently, the government has been experimenting with species-specific total allowable catch (TAC) quotas, including a suite of supporting monitoring and enforcement mechanisms, such as catch logbooks, vessel inspections and at-sea observers (Huang & He, 2019). The revised 2000 Fisheries Law mandated TAC management in China, but inadequate scientific information and lack of supporting management infrastructure have been barriers to implementation (Su et al., 2020). Experimental implementation of TAC pilots began in 2017, providing provincial governments with experimental opportunities to improve weak data collection, catch monitoring and enforcement capacity at a relatively small scale. To date, pilot tests have been limited to nearshore invertebrates and juvenile Japanese anchovy, and their effectiveness still is under evaluation.

A more sweeping policy has been annual summer moratoria, applied for different durations in different seas since 1995, whereby commercial fishing is closed to relieve fishing pressure and protect spawning opportunities. After some years of pilot testing at the provincial level, especially in Zhejiang Province, summer moratoria began at the national level with two or three-month closures in the Bohai Sea, East China Sea and Yellow Sea in 1995, and later extended to the South China Sea in 1999. Today, the moratorium begins May 1 for all waters and commercial fishing gears except angling, and extends to over four months in some regions. The closures appear to help protect certain spawning stocks and juveniles for the duration of the closure, but short-term protections are usually offset by high pulses of fishing effort immediately after each closure, dissipating benefits within months (Shen & Heino, 2014). The policy trend of lengthening the moratoria and expanding its scope over the years suggests disproportionate reliance on this management measure, amidst the absence or ineffectiveness of others. Moreover, by idling the nation's 200,000 commercial fishing vessels for several months each year, the policy has socioeconomic impacts which make it difficult for fishery workers to find alternative employment opportunities during the extended closure months (Cao et al., 2017).

Altogether, despite zero, and later negative, growth policies since 1999 and an expanding portfolio of fishery management tools, the total number of fishermen and the fleet's harvesting capacity have increased steadily over time (Fig. 3.2). The nation's fishing fleet and number of fishers tripled in size between 1990–2010 (Shen & Heino, 2014). The total number of fishing vessels has decreased in recent years, but the cumulative total engine power of the fleet continues to increase, suggesting technological advances among existing vessels and replacement of older, smaller vessels with larger ones with greater fishing capacity that require larger fishing crews to operate. Much of this capacity expansion appears directed at distant water fisheries, as domestic catch has remained relatively stable since 2000. While distant water catch represents a significant and growing percentage of total marine catch at 18%, high fishing pressure in China's domestic waters by overcapitalized fleets remains a pressing challenge.

The government's capacity reduction goals have been complicated by policies to subsidize marine fishing. In 2006, the national government began subsidizing agricultural production, including fisheries. In China, fisheries subsidies have taken the form of direct vessel purchases, fishing employment and vessel insurance, tax deductions and fuel subsidies (Cao et al., 2017). Policies to reduce fishing costs incentivize more fishing capacity and effort, illustrating the government's difficulty in balancing potentially competing goals of resource conservation with employment and income stability. Nevertheless, its spending on subsidies has increased dramatically over a short time, growing from US$810 million in 2006 to US$6.5 billion by 2013, with nearly all allocated to fuel subsidies (Mallory, 2016; Zhang & Wu, 2017). Despite capacity reduction goals, a significant proportion of fuel subsidies are allocated to domestic water fleets. Overall, government subsidy expenditures amount to 20% or more of the value of China's total marine catch (Mallory, 2013).

The government's challenge is complicated even more by the prevalence of illegal fishing gears and unlicensed vessels. For example, in Zhejiang Province, the number of unlicensed and illegal fishing vessels peaked at 13,000 in 2013, when total catch in the East China Sea reached three MT, despite the government having set a target for catch at two MT (unpublished data from Zhejiang Ocean and Fishery Bureau). Inadequate numbers of enforcement personnel relative to the large number of fishers and vessels in China make it difficult to adequately monitor the fishing activity and catch of legally permitted vessels, let alone detect and punish illegal vessels.

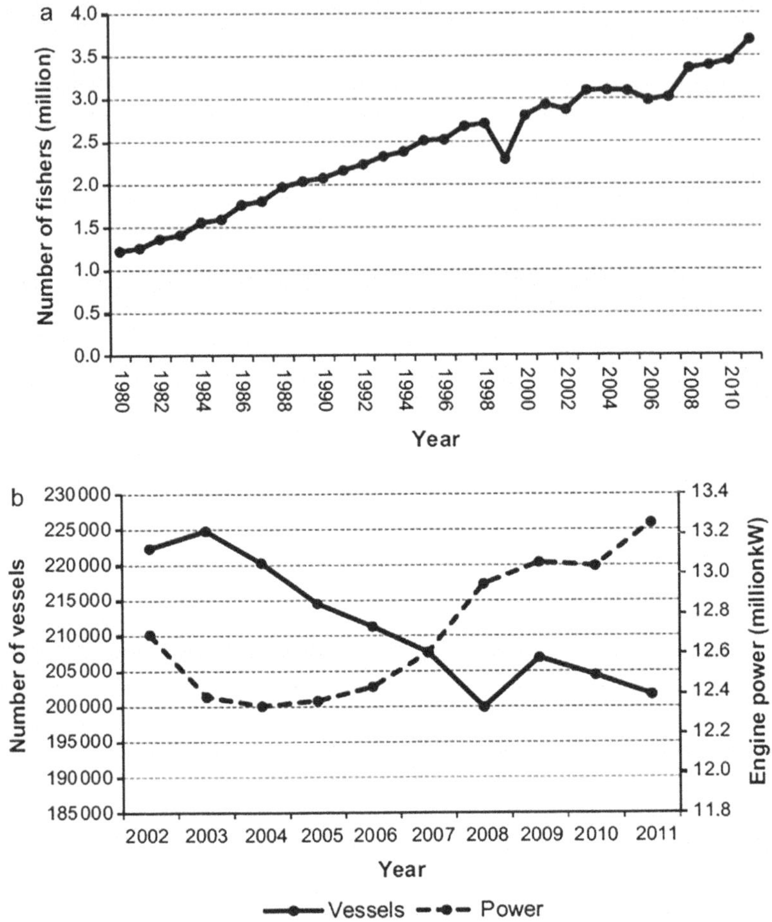

Fig. 3.2 Changes in the total number of fishers (**a**) and Fishing power in terms of both the number and engine power of vessels (**b**) in China through time (*Source* Shen & Heino, 2014)

3.4 Social and Economic Factors

Addressing overcapacity in fishing fleets is made more difficult by important and powerful social and economic forces. Pressure to maximize near-term economic gains, especially for impoverished rural communities with few other livelihood options, can cause long-term productivity and stability to be compromised in favor of short-term fishing yields. Similar tradeoffs can be made when an unregulated industry sees an opportunity to capitalize on price trends or establish a strong position in supply chains. These tradeoffs eventually lead to ecological degradation, economic loss and social hardship. Unfortunately, these adverse outcomes encourage fishermen to fish even harder to offset their losses, which causes further degradation, creating a destructive cycle.

Arguably, the solution to fishery recovery and stability is straightforward on the surface: simply reduce fishing pressure to allow stock to recover, then fish the recovered stock at a conservative rate. With a large fish population, even modest fishing pressure can generate high yields and profits because the underlying resource is abundant and catch efficiency is high. Indeed, a recent global analysis suggests that, despite the high number of overexploited and collapsed stocks worldwide, recovery and sustainable management can lead to even higher yields and profits in the future (Costello et al., 2016). However, very real economic and social pressures often prevent governments from undertaking the actions needed to recover fish stocks. If aggressive action toward recovery is taken, fishing communities could undergo a period of hardship, especially if no other actions are taken to address these communities' economic needs. This may be a short-term problem if stock recovery is rapid, although any period of time without food or income could be devastating to fishing communities with few other viable economic options. Furthermore, overcapacity that contributed to fishery declines in the first place might require some participants exiting the fishery in order to achieve and maintain sustainability going forward.

Subsidies generally compromise the economic and ecological viability of fisheries (World Bank, 2009), although time-limited subsidies as investments in keeping fishing communities viable during periods of management reform and stock recovery can be an effective strategy. To date, this has not been the nature of fishery subsidies in China; however, in 2015, the Ministry of Finance (MOF) and MARA concluded that the extensive fishing vessel fuel subsidy policy, in place since 2006, was at odds with policies to reduce the size of fishing fleets and constrain harvests. Therefore, the central government announced that the fuel subsidy would be reduced to 40% of the 2014 peak by 2019 (MOF & MOA, 2015). The practical impact of this policy change remains unclear.

Relocation of fishermen to other livelihoods is perhaps the most important in a series of actions through which Zhejiang Province has worked to restore the East China Sea fishing grounds. This is a considerable challenge, as 75% of China's fishermen come from families that have been fishing for generations and have few alternative income opportunities (Su et al., 2020). Some displaced members of the commercial fleet have been able to remain employed in fisheries through new opportunities in the growing recreational fishing sector, while others have joined the distant-water fleet. However, like other industries in the coastal zone, over time many fisheries in China have employed larger numbers of migrant laborers; for example, it is estimated that 30% of fishermen in Zhejiang Province are from other provinces. Generally, welfare policies only apply to legal residents of a province, which means that relocation programs do not address the social and economic needs of the whole fleet, and welfare programs are often not applicable to middle aged or elderly fishermen who might still need income to support their families.

3.5 Scientific and Technical Challenges

Fisheries lacking scientific assessments are more likely to decline and collapse (Costello et al., 2012), which means that the perceived poor status of many Chinese marine fisheries stocks may likely be accurate. That China's nationwide fishing fleet continued to grow, even as total catch remained largely unchanged, suggests that stocks are in fact dwindling, as it is likely that only through increasing catch capability attributable to the growth in fleet have total catch levels been maintained. However, limited stock assessments and issues with data quality and availability preclude rigorous evaluation of the full extent of China's overfishing challenges.

Typically, sound science and sound policy reinforce one another: policies can drive demand for robust science, and science can improve the efficacy of policy. For example, mandates to define clear goals, targets, limits and requirements for risk-averse management, require scientific information and can provide government and academic scientists a clear focus for their work, as do policies that require consideration of appropriate spatial scales in management and awareness of ecosystem factors that affect, and are affected by, fishery management decisions. The greater dedication to management in China has led to continued strengthening of national policies, but this progress has not kept pace with the magnitude of overcapacity and environmental degradation that have accumulated since China's fishing industry began to substantially develop in the mid-twentieth century; one of the important reasons is that there is a lack of complementarity between fishery science development and fishery policy improvement.

Despite these problems, China has the benefit of a strong scientific community to help meet its fishery management challenges. Academic institutions and research institutes in China are highly regarded across the globe, and fishery researchers in China have impressive publication records. Yet, this capacity is not fully utilized due to structural barriers that unnecessarily compartmentalize individuals and institutions. Transparency and accessibility of fishery data are extremely limited; consequently, universities and provincial fishery research institutes often do not have full access to data collected by the larger regional fishery research institutes, which would enable them to do more innovative research and complete more and better stock assessments. Data sharing can also pave the way for increased collaboration, which can improve the overall quality and quantity of technical analyses supporting fishery management.

Of course, even if these structural barriers are removed, and data begins to flow more freely and productive collaborations grow across institutions, fishery science in China will only be as reliable as the data feeding into models and analyses. In particular, accurate data on total catch of each stock, as well as higher resolution data, such as catch by different gears, in different seasons or areas, or at different life stages of exploited organisms, are important in any fishery management program (Boenish et al., 2020; Kritzer, 2020). Unfortunately, inadequate catch monitoring also means that these data are often of low quality, or altogether nonexistent. Improving monitoring is arguably the single most important challenge facing fishery management in

China. In addition, consideration and application of data-limited stock assessment methods that can utilize existing or near-term data can support stronger management in the interim period while more sophisticated data and information is being collected over the longer term.

The importance of monitoring lies in not only its scientific benefits, but its value for enforcement and shaping fishing behavior. As monitoring improves in a fishery, the range and effectiveness of management strategies that can be adopted also improve. For example, global experience shows that the implementation of TAC management, especially with quota allocated to individuals, vessels, communities, cooperatives or other entities, can be a highly effective strategy for fishery management (Costello et al., 2008). However, without effective monitoring, these approaches can incentivize unreported discarding of catch at sea, which results in management targets being missed and scientific assessments based on faulty data. In the absence of the monitoring necessary to allow adoption of output controls, China has instead relied primarily on input controls. Some of these have had important benefits, but they cannot address all of China's management needs, and often introduce inefficiencies that exacerbate socioeconomic strains on fishing fleets and communities. Encouragingly, China has initiated a series of pilot projects in TAC management at the provincial level, several of which are prioritizing catch monitoring as a core component.

The catch of fisheries in China is highly multispecies due to the nature of China's ecosystem, the use of unselective gears, and the absence of limits on the harvest of individual species. Chinese consumers also have a broad palate for seafood, so there is less market pressure on fishing fleets to be selective in their catch. Diversified catch portfolios can have ecological benefits by spreading impacts more evenly across species and trophic levels (Garcia et al., 2012). However, sustainably managing such diversity only increases the need for collaborations across scientific institutions and improved data streams that greater catch monitoring can provide. Furthermore, new scientific tools and management strategies will need to be developed to avoid ecosystem-scale degradation. Testing and implementation of these new monitoring, scientific and management approaches will be more effective through close collaboration with the fishing industry, in order to ensure buy-in and benefit from their unique knowledge of the fishery.

3.6 Summary and Conclusions

As the dominant fishing nation in the world, with unparalleled levels of catch and international reach through fishing operations and supply chains, fisheries management in China has environmental, economic and social implications at the global scale. The history of fisheries in China has followed that of many nations, with rapid and unmanaged escalation that led to severe resource depletion, exacerbated by pollution, development and other environmental impacts, with management efforts lagging and struggling to catch up. Despite these challenges, environmental policy mandates in China are growing stronger, with tangible progress underway in the fisheries sector. A world-class scientific infrastructure stands ready to support these policy transformations, although it will need to be more effectively coordinated and mobilized. Climate change (see Chap. 7) is presenting new challenges, but continued technological innovation enables more rapid and effective tracking of, and response to, these impacts. Increasing industry involvement will be essential to ensuring that the integration of science, management and technology ultimately deliver economic and social benefits on the water in ways that protect China's unique marine natural heritage.

Chapter 4
Habitat and Biodiversity

Ling Cao, Zihan Wang, Shuo Cheng, Chunyu Liu, Ziyu Jiang, and Hui Liu

China has recognized the importance of marine habitats and started to conserve them in a much more dedicated manner. As discussed further in Sect. 4.4 of this chapter, this has been compelled by stronger policy imperatives, backed by government support for increased monitoring, research, and protection and restoration projects in many important coastal and ocean areas. Despite this progress, major challenges remain and overall management effectiveness is limited. The key challenges lie in continued deficiencies in the policy architecture—particularly in the establishment of a strong, clear and comprehensive governance system—and an inadequate technical foundation that precludes both effective implementation of existing policies and creation of stronger policies.

L. Cao (✉) · S. Cheng
Xiamen University, Xiamen, China
e-mail: caoling@xmu.edu.cn

S. Cheng
e-mail: chengshuo@xmu.edu.cn

Z. Wang · C. Liu · Z. Jiang
Shanghai Jiao Tong University, Shanghai, China
e-mail: ilxlddove@sjtu.edu.cn

C. Liu
e-mail: chunyu_liu@sjtu.edu.cn

Z. Jiang
e-mail: markjiang@sjtu.edu.cn

H. Liu
Yellow Sea Fisheries Research Institute, Chinese Academy of Fishery Sciences, Qingdao, China
e-mail: liuhui@ysfri.ac.cn

H. Liu et al. (eds.), *Conservation and Sustainable Use of Living Marine Resources and Biodiversity*, Global Ocean Governance and Ecological Civilization,
https://doi.org/10.1007/978-981-19-4790-2_4

4.1 Coastal and Marine Ecosystem Health in China

China has a vast sea area and a long coastline of more than 3.2×10^4 km, including a land coastline of 1.8×10^4 km and an island coastline of 1.4×10^4 km (Zhou et al., 2016). China spans different temperature zones and different topographic and geological regions, thus has high biodiversity and diverse coastal and marine ecosystems, including estuaries, tidal flats, mangroves, coral reefs, seagrass beds and so on. As of 2013, coastal wetlands in China covered a total of 5.80×10^4 km^2, accounting for 10.85% of the total wetland area in China (Fig. 4.1). These are mainly distributed in 11 provinces and cities, of Liaoning, Hebei, Tianjin, Shandong, Jiangsu, Shanghai, Zhejiang, Fujian, Guangdong, Guangxi and Hainan, among which the area of coastal wetlands of Shandong and Guangdong provinces exceed any other province and together accounts for 37.55% of the total coastal wetlands in China (Fig. 4.2) (Sun et al., 2015). China's coastal wetlands could be divided into two groupings according to their geography. To the north of the Hangzhou Bay, besides the rocky shores in southeastern Liaodong Peninsula and northern Shandong Peninsula, the rest of the shorelines (incl. the rim of Bohai Sea and inshore-coastal wetlands of Jiangsu Province) are sandy or silty tidal flats. Coasts to the south of the Hangzhou Bay are generally rocky, including the Qiantang River–Hangzhou Bay, the Wenzhou Estuary, the Jin River Estuary–Quanzhou Bay, the Pearl River Delta and the Beibu Gulf. Furthermore, coral reefs are mainly distributed in the coastal areas of Xisha Islands, Nansha Islands, Hainan Island and Taiwan Island, and natural mangrove forests are mainly distributed along the coasts from Hainan Province to Fujian Province and on the west coast of Taiwan (Niu et al., 2009).

Coastal wetlands are the essential ecological barrier for sustainable economic and social development. Coastal wetland ecosystems have massive environmental functions and ecological benefits in many aspects, such as water purification, climate modulation, flood and hazard control, storm surge and typhoon mitigation, providing

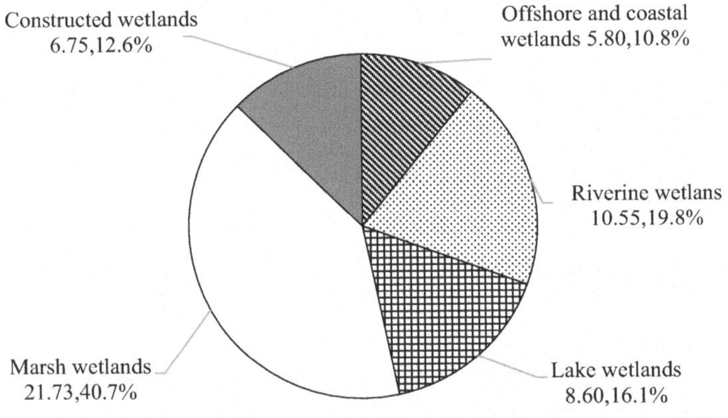

Fig. 4.1 Area and percentage of different types of wetlands in China (10^4 km^2) (*Data Source* China Forestry Statistical Yearbook 2017)

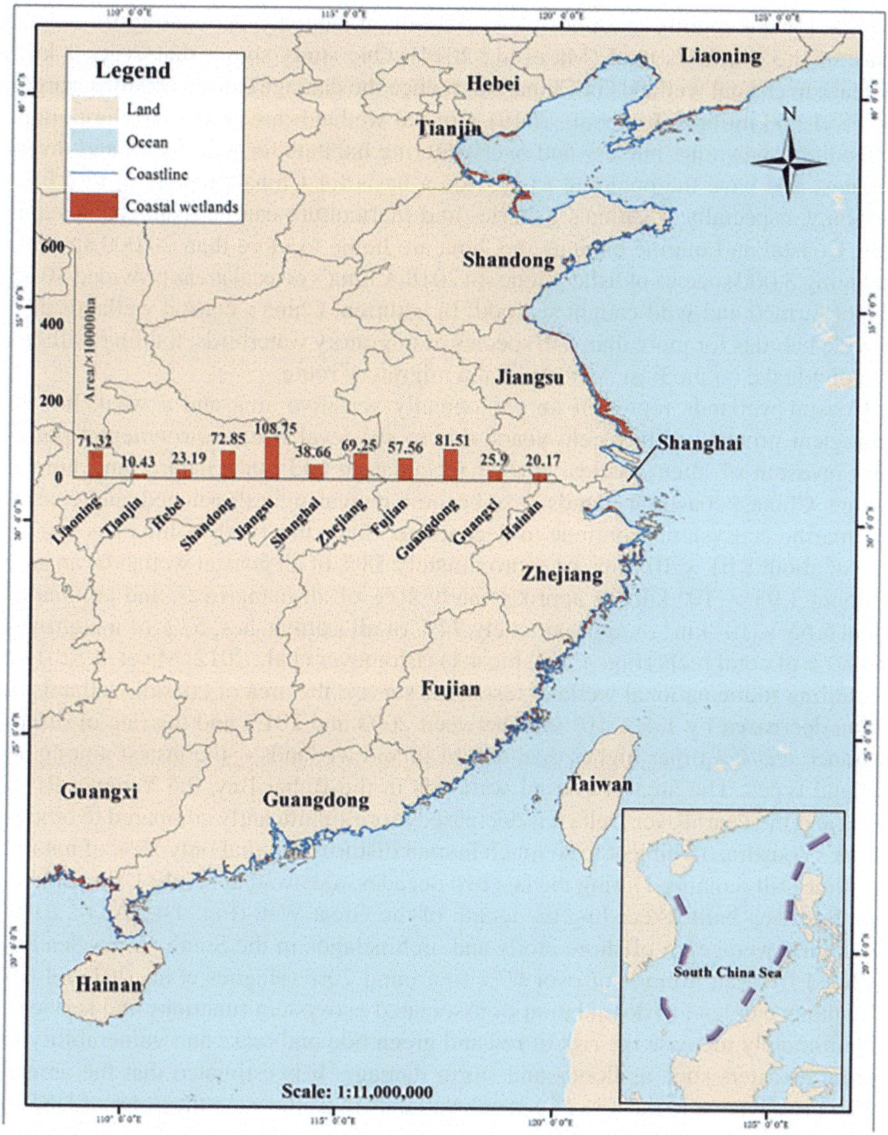

Fig. 4.2 Distribution of China's coastal wetlands in 2016 (Liu et al., 2020)

scenic and recreational value, providing buffers from natural disasters for towns and villages along the coast, thereby protecting peoples' lives and property. Mangroves, for example, not only have the general functions of a forest ecosystem, but also can protect against natural disasters such as storm surges and tsunamis, prevent coastal erosion, and increase carbon sinks. Although coastal wetlands account for less than 1% of the area of all ecosystems in China, the values of ecosystem services they provide for the country amount to US$200 billion annually, accounting for a 16% share of the country's total (Ma et al., 2014). One study shows that every 1 km^2 increase in coastal wetlands in China will reduce the damage caused by storm surges by US$12.68 million (Liu et al., 2019). Coastal wetlands are particularly important as feeding, spawning, nursery and overwintering habitats for wild fishes and invertebrates, and have the potential to provide a basis for China's transition to a blue economy, especially if China's fisheries and mariculture can be managed sustainably. Coastal and marine habitats in China are home to more than 20,000 species, including 3,000 species of fishes alone. In 2018, China's coastal areas provided 30.75 MT of farmed and wild-caught seafood. In addition, China's coastal wetlands also provide habitats for more than 240 species of migratory waterbirds, totaling millions of individuals, on the East Asia-Australia migratory route.

Coastal wetlands represent an ecologically sensitive area and a weak link in ecological protection. In recent years, due to sea level rise, environmental pollution, invasion of alien species, coastal reclamation and other natural and human factors, China's coastal wetlands have become increasingly threatened, and coastal and marine ecosystems continue to degenerate. Since the 1950s, China has lost an area of about 8.01×10^4 km^2 or approximately 58% of all coastal wetlands, an area of about 1.93×10^4 km^2 or approximately 40% of all mangroves, and an area of about 5.65×10^3 km^2 or approximately 74% of all saltmarshes, 57% of mangroves and 80% of coral reefs (Fig. 4.3, Table 4.1) (Blomeyer et al., 2012; Ma et al., 2014). According to the national wetland resources survey, the area of coastal wetlands in China decreased by 1.4×10^4 km^2 between 2003 and 2013, and the rate of disappearance was 2.4 times higher than that of inland wetlands,—the fastest among all wetland types. The area of coastal wetlands in the Bohai Bay, the Yangtze River Delta and the Pearl River Delta has decreased more significantly compared to others. China's coastline is subject to so much human disturbance, that only 30% of natural coastline still remains. During the last two decades, a seawall at nearly 1.1×10^4 km long has been built, exceeding the length of the Great Wall (Fig. 4.4). As of 2010, the coral coverage on offshore atolls and archipelagos in the South China Sea has declined from an average of over 60% to around 20% (Hughes et al., 2013). Loss of habitats can lead to degradation of associated ecosystem functions and services, and ultimately increase the risk of red and green tide outbreaks and vulnerability to natural disasters such as floods and storm damage. It is estimated that the annual economic cost due to the loss of coastal wetlands in China is on the order of US$46 billion (An et al., 2007).

Facts have proved the above inference. In the last decade, red tides occurred 55 times per year on average in China's coastal waters, with East China Sea more seriously hit by more cumulative occurrences and affected areas (Table 4.2). The green tides, which has even greater economic impact, mainly occurred in the Yellow Sea (Fig. 4.5), causing direct economic losses of RMB 1.3 billion in 2008 (Qiu, 2011). Some marine ecosystems, most notably the Bohai Sea and northern Yellow Sea, have been severely degraded and become seasonally hypoxic (Gao et al., 2014; Zhai et al., 2019). Significant eutrophication has also occurred in Liaodong Bay, Bohai Bay, Jiaozhou Bay, Yangtze River Estuary, Hangzhou Bay, Minjiang Estuary

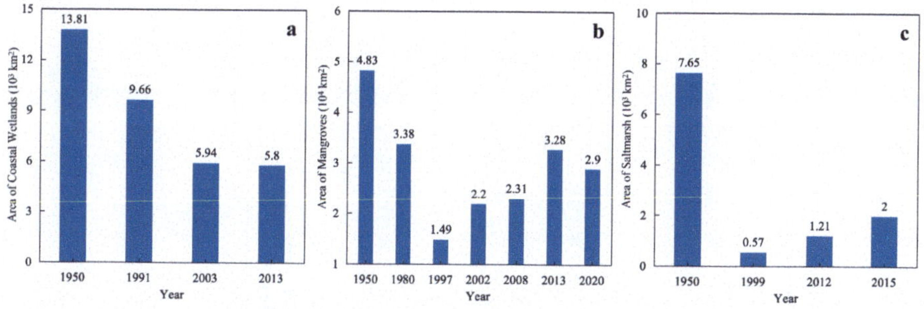

Fig. 4.3 Areas of **a** coastal wetlands, **b** mangroves and **c** saltmarsh in China

Table 4.1 The area of typical coastal habitats in China (10^4 km)

Habitats	Time span	Area	Source
Mangroves	1950–2020	Figure 4.3b	Qiu (2011); Sun et al. (2015); Fu et al. (2009)
Saltmarsh	1950–2015	Figure 4.3c	Shi et al. (1995); Luo (2019)
Seagrasses	2013	8.7651	Zheng et al. (2013)
Coral reefs	2013	6.4608	*China Forestry Statistical Yearbook 2013*

and Pearl River Estuary, together compromising survival of fishes and other living marine resources.

4.2 Biodiversity Loss

China is endowed with rich living marine resources, with more than 28,000 species of marine organisms recorded, accounting for 12% to 15% of the total number of recorded LMR worldwide (Huang et al., 2012). According to the results of marine biodiversity monitoring conducted in 2018, 1572 species of macrobenthos, 686 species of zooplankton, 718 species of phytoplankton, and 85 species of reef-building corals live in the China seas (Table 4.3, Fig. 4.6). The coral reef ecosystems in the China seas exhibit high biodiversity. More than 200 species of scleractinian corals live in the northern South China Sea, about two-thirds of the total species worldwide. They are distributed mainly along the southern coasts of Hainan and Taiwan, and around Dongsha, Xisha, and Nansha islands (Liu, 2013).

Although marine biodiversity in China has generally been stable over the last decade (Fig. 4.6), it has shown a significant decline compared to the mid-twentieth century. On the one hand, China's marine trophic level index has fluctuated and declined (Du et al., 2014). Between 1974 and 1997, this index has declined by 7.18%, and rose steadily to 3.34 over the next 15 years; however, this was below the relatively higher value of 3.45 in 1974 and before, and also lower than the global average (Fig. 4.7). The decline in the marine trophic index indicates that China's

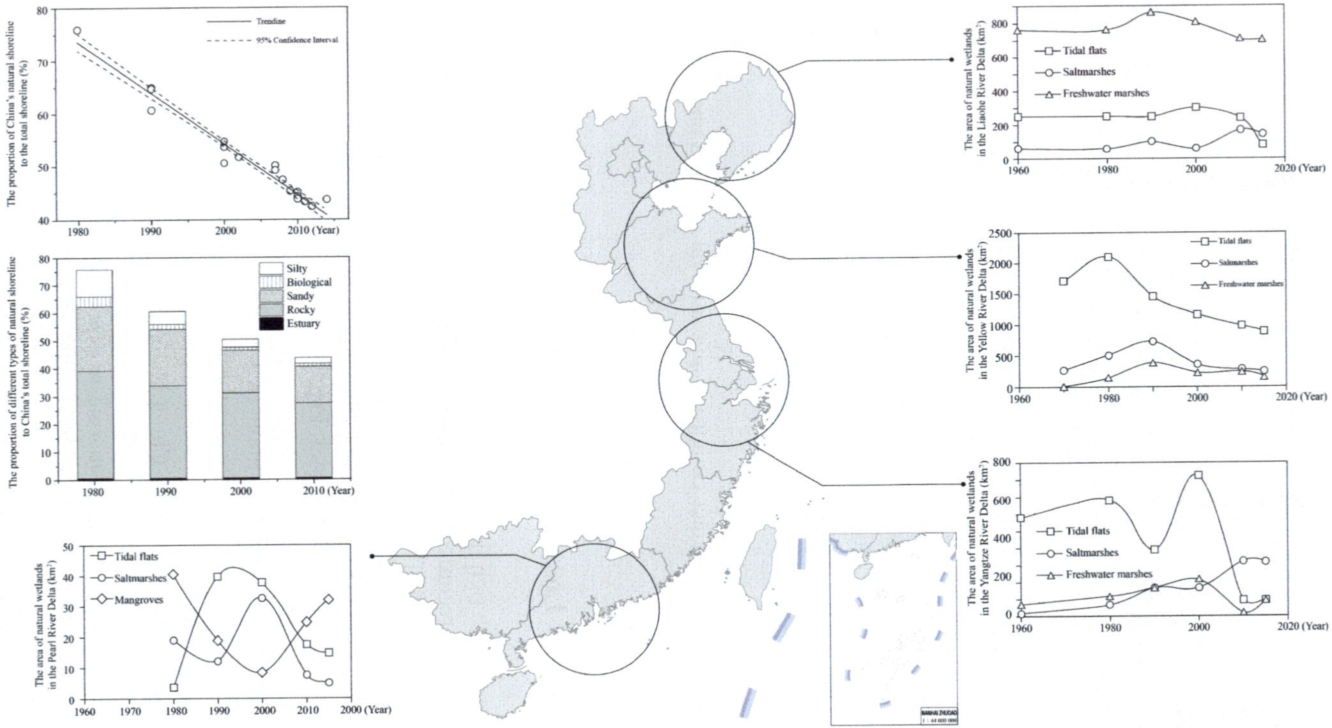

Fig. 4.4 Changes in the length of China's natural shoreline and the area of natural wetlands in the four major deltas

Table 4.2 Red tides occurrence in the China seas in 2019

Area	Number of red tides	Cumulative area of red tides (km^2)
Bohai	2	0.28
Yellow Sea	2	5
East China Sea	31	1974
South China Sea	3	12
Total	38	1991.28

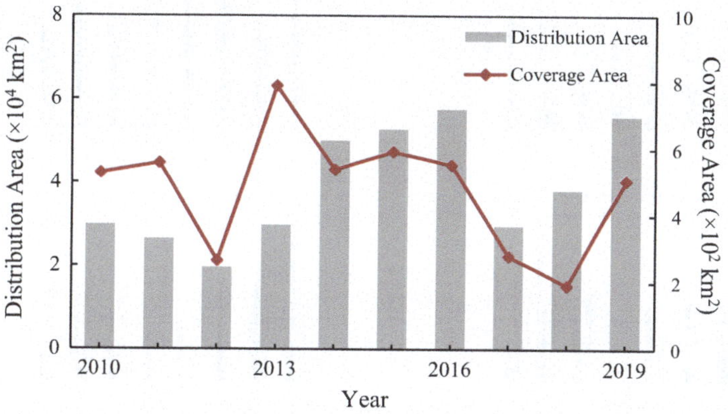

Fig. 4.5 The occurrence of *Enteromorpha* sp. green tides in Yellow Sea from 2010 to 2019 (*Data Source* National bulletin on marine disaster (2013–2017))

Table 4.3 Number of biological species in waters under Chinese jurisdiction in 2018

Area	Phytoplankton	Zooplankton	Macrobenthos
Bohai	171	85	286
Yellow Sea	212	113	305
East China Sea	468	439	699
South China Sea	486	505	972

marine biodiversity and marine ecosystem integrity have declined, with the main catches changing from long-lived, high-trophic benthic fish to short-lived, low trophic invertebrates and pelagic fish. On the other hand, the abundance and distribution of species have changed significantly in some regions. For example, in Jiaozhou Bay on coastal Yellow Sea, a total of 34 species of benthic invertebrates dominated by Mollusca and Crustacea were found in a piece of sandy-muddy tidal flat on the east coast between 1935 and 1936. The species composition and abundance in the area increased over the following decades, probably due to different sampling methods, as 141 species including 52 species of Crustacea, 41 species of Polychaeta, 40 species of Mollusca and 40 species of Echinoderms were recorded between 1963 and 1964.

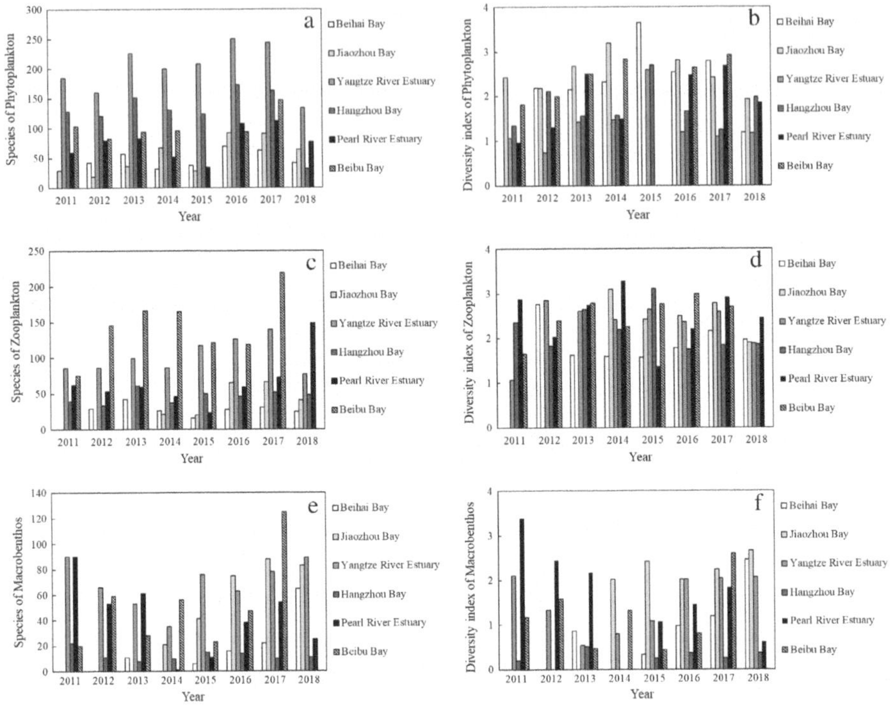

Fig. 4.6 Status of China's marine biodiversity from 2011 to 2018 (*Data Source* China's Inshore Marine Environment Quality Bulletin (2011–2017) and China's Marine Ecological Condition Bulletin 2018)

The biodiversity then changed greatly after the 1970s because of heavy industrial pollution and the disruption of intertidal environment due to coastal exploitation; unfortunately, no live benthic animals were observed in the same tidal flat after 1989 (Liu et al., 1983; Briggs, 1974; Schram, 2010). According to incomplete statistics, the number of benthic species in the whole tidal flats of Jiaozhou Bay has decreased from 120 in the 1960s to about 20 recently. In addition, the population of the dominant cold-water benthic species (such as hermit crab) in the central Yellow Sea, which is the most unique among China's offshore biota, has also been decreasing over the past 50 years.

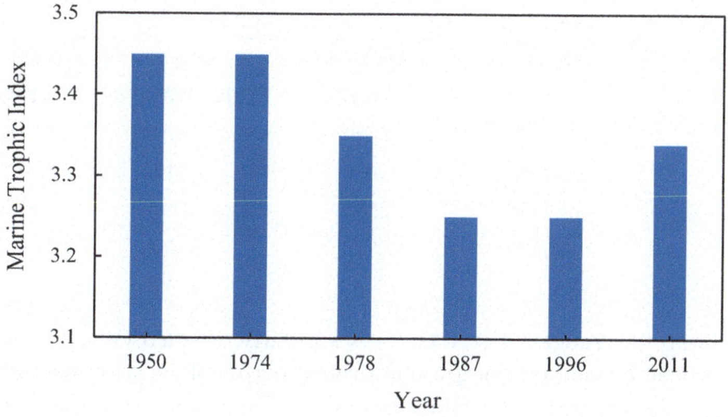

Fig. 4.7 Change in marine trophic index in China from 1950 to 2011 (*Data Source* Du et al. 2014)

4.3 Causes of Marine Habitat Destruction and Biodiversity Loss

China's coastal and marine ecosystems are under great threat due to both human and natural factors. In order to ensure the sustainable use of coastal living marine resources, we should systematically analyze the main causes of and challenges to marine habitat destruction and biodiversity loss, with a further focus on the scientific management of marine habitats and biodiversity.

In China, coastal regions have seen the most active economic and social development. The coastal provinces and cities have accommodated 40% of the population and over 50% of large cities, and have generated 60% of GDP for China. The proportion of China's marine economy in GDP has increased from 1.9% in 1996 to 9.3% in 2018 (He et al., 2014). This rapid development of coastal areas has led to serious pollution, which put tremendous pressure on coastal habitats and the marine ecological environment. The sources of pollution mainly came from domestic and industrial wastes, excessive discharge of agricultural fertilizers and pesticides, and large-scale aquaculture. Wastewater and solid wastes from various sources significantly pollute coastal regions, either through direct discharge or via riverine inputs to coastal seas (Table 4.4). Since the early 1990s, coastal pollutant inputs have increased by about 1.1×10^9 t per year, and more than 90% of these are carried by rivers (Xu and Zhang, 2007). As a result, water contamination levels at the riverine–marine interface often exceed the limits set by the *Integrated Wastewater Discharge Standard (GB8978-1996)* or the *Discharge Standard of Pollutants for Municipal Wastewater Treatment plant (GB18918-2002)*, and the content of main pollutants (COD, nutrients, petroleum, heavy metals and arsenic) in the Yangtze River Estuary and the Pearl River Estuary are significantly higher than other typical estuaries.

In addition, more than 60% of the coastal wetlands along the Bohai Sea and Pearl River Delta are affected by high or medium level pollutant input, while 44.7%

Table 4.4 Status of eutrophication of the China seas in 2018 (km^2)

Area	Mild	Moderate	Severe	Total
Bohai	3220	660	370	4250
Yellow Sea	9240	4630	310	14,180
East China Sea	7960	10,030	11,740	29,730
South China Sea	4170	2590	1760	8520
Total	24,590	17,910	14,180	56,680

wetlands along the Yangtze River Delta is similarly affected (Wang et al., 2013). Since the 1960s, the demand for grain has risen sharply along with the rapid growth of coastal population, stimulating the conversion of traditional land types to cropland and mariculture. Therefore, agriculture has become an important source of pollution for coastal wetlands. In some regions where agriculture is extensive, the nutrient content in coastal waters has increased by more than ten-fold over the past decades, of which more than 50% came from scattered agricultural activities (Cao et al., 2003). For example, eutrophication and overgrowth of benthic algae has negatively affected the coastal waters of the Jiulong River in Fujian Province since the mid-1980s, where 60.9% of total nitrogen came directly from agricultural activities. Water pollution may affect the reproduction and recruitment of offshore organisms, and increase the vulnerability of marine organisms to diseases, thereby reducing marine biodiversity; it may also affect the productivity of coastal wetlands and the quality of coastal habitats, by interfering with the local vegetation.

Excessive coastal reclamation is one of the main pressures on China's coastal and marine ecosystems. Since China's reform and opening up, the coastal regions have entered a stage of rapid urbanization. Rapid economic and population growth has brought about pressure on urban living space. Besides, a large number of ports and docks were to be built to cope with the huge trade throughput, and more land area were needed for building factories so as to adapt to the trend of transferring China's steel, petrochemical, shipbuilding, thermal power, nuclear power and other heavy industries to the coast. The huge demand for land as such generated, were mainly met by reclamation of coastal waters and wetlands (Fig. 4.8). From 1985 to 2010, a total of 7547 km^2 of coastal wetlands were reclaimed in China, at an annual growth rate of 5.9%; and between 2009–2014, about 500 km^2 of coastal areas were reclaimed (He et al., 2014). Reclaimed coastal wetlands were mainly for mariculture, agriculture and industrial uses (Fig. 4.9) (Cao et al., 2017).

Rich in biodiversity, the Yellow Sea region contains important fishing grounds and is an important stopover for migratory birds. Since the early 1980s the region has lost 35% of intertidal habitat area due to reclamation, especially along the coasts of Jiangsu Province at the southern extent. Much of the intertidal zones and shallow waters, especially in the Bohai Sea, is now occupied by mariculture ponds and cages. The severe impact of unrestrained coastal exploitation is presented as follows: (1) coastal erosion or siltation; (2) insufficient gulf water exchange capacity and deterioration in water quality; (3) the degradation and loss of wetland function, and structure

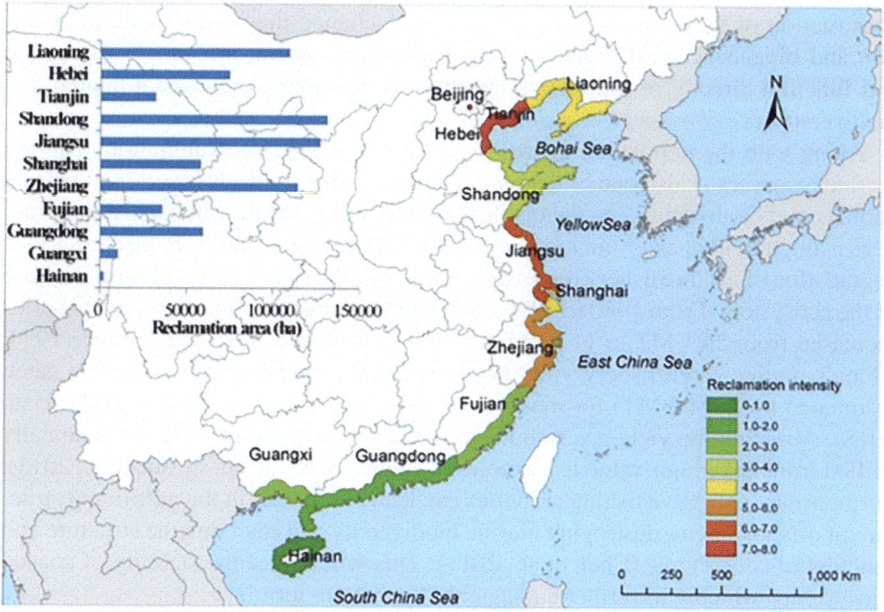

Fig. 4.8 Area and intensity of coastal reclamation in China from 1985 to 2010 (Tian et al., 2016)

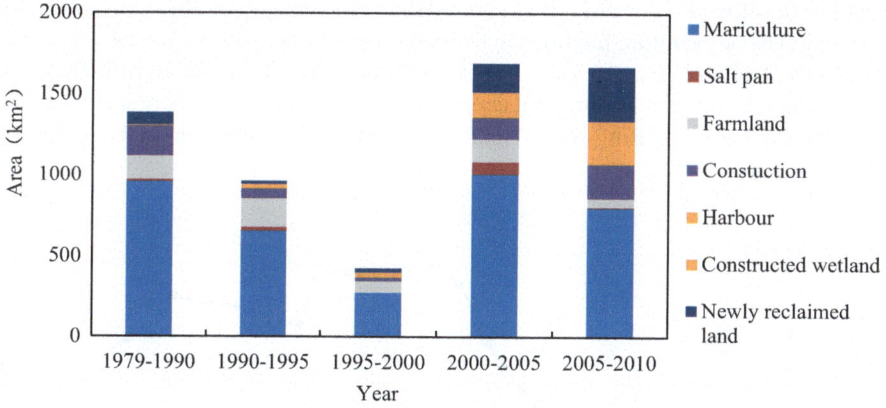

Fig. 4.9 Use of reclaimed coastal wetlands in China from 1979 to 2010

of coastal and marine ecosystem; and (4) destroyed landscape resources. As of 2012, more than 30% of sandy native coastal lands and 60% of landscapes such as offshore sand bars and coastal lagoons were damaged, respectively (He et al., 2014). Land reclamation will destroy coastal wetlands, which is the natural ecological barrier helping to prevent marine disasters such as storm surges in the sea-land interface; as a result, sea dykes will need to be built in defense of such disasters. However, the construction of these engineering projects may change the original natural evolution and biogeochemical process of the coastal ecosystem, affecting its structure and function directly or indirectly, and thereby destroying the coastal habitat and biodiversity.

Along with the population growth in coastal regions, overexploitation of living marine resources is also one of the main factors leading to the destruction of marine habitats and biodiversity. On the one hand, overfishing and destructive fishing practices using trawling, exterminating nets and pumping boats, may both lead to serious degradation of inshore fishery resources and change the structure, function and nature of the ecosystem. From 1980 to 1996, the volume of domestic marine catch fisheries increased from 2.81 MT to 11.53 MT, with a growth rate of 310.32%. Since 1994, China's marine catch fishery yield has greatly exceeded the total allowable catch (estimated to be 9.00 MT) for many consecutive years (Fig. 4.10) (Lu, 2005; Han, 2018). Although the yield has stabilized in recent years, the catch has dramatically shifted from large high-value fish species to small low-value ones (Sun et al., 2015). Furthermore, excessive fishing activities can lead to changes in the nutritional structure of offshore areas, destroying marine biodiversity and changing the structure and function of ecosystems (Chen et al., 2011). This will reduce the stability of coastal ecosystems, which will further damage biodiversity in turn.

On the other hand, the excessive development of large-scale mariculture also puts pressure on coastal and marine ecosystems and reduces the abundance of wild fish and benthic organisms. In recent years, the aquaculture industry in China's coastal areas has developed vigorously. It's reported that a total length of 362.2 km coastline is occupied by aquaculture facilities in Hainan Province, accounting for 19.9% of the island's total natural coastline of 1822.8 km (Zhou et al., 2016). The development of the mariculture industry has brought economic benefits, but it will encroach upon the natural coastal zone, causing coastal habitat degradation and weakening the aesthetic value of the landscape.

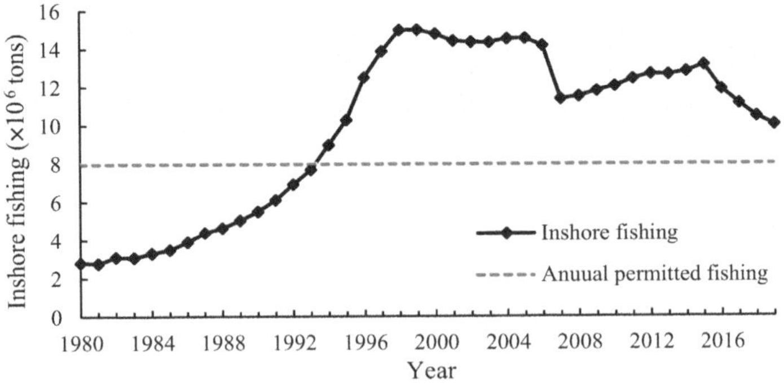

Fig. 4.10 Variations in total yield of offshore catch fisheries in China, 1980 ~ 2018. (*Data Source* Lu 2005; FOB 1980–2018)

At the same time, the invasion of non-native species and coastal erosion are non-negligible causes of the degradation of wetland ecosystems. Escape of alien species introduced by mariculture may also have a negative impact on marine ecology. Since the 1950s, China has introduced hundreds of new aquaculture species from abroad, many of which are common mariculture species such as red abalone *Haliotis rufescens*, Pacific oyster *Crassostrea gigas*, Gulf scallop *Argopecten irridiance*, giant seaweed *Macrocystis pyrifera*, and turbot *Scophthalmus maximus*, which will enter the natural coastal ecosystem because of natural disasters and negligent management. These escaped species may not only directly eliminate native species due to their high adaptive capacity and lack of natural predators, but also cause genetic contamination by interbreeding with native species. So far, about 26 kinds of economically important marine species (finfish, mollusks, and shrimp) and three species of aquatic plants have been introduced to the coastal regions of China. These aquatic plants (*Alternanthera philoxeroides*, cordgrass *Spartina alterniflora* and *Eichhornia crassipes*) are already identified as invasive species, causing an annual economic loss of about US$20 million. *S. alterniflora*, which was introduced to China in 1979 for shoreline protection, was included in China's first list of invasive alien species in 2003. Due to its strong reproductive capacity and tolerance to salinity and inundation, *S. alterniflora* had spread to ten coastal provinces, and its coverage in Jiangsu, Shanghai, Zhejiang and Fujian together made up 94% of its distributional area in China (Fig. 4.11) (Zuo et al., 2012), causing serious degradation of China's coastal wetlands, especially tidal flats to the south of the Bohai Bay.

Since the 1960s, China's coasts have experienced intensified erosion, which causes irreversible changes to coastal wetlands and seriously affects the stability of China's coastal ecosystems. More than one-third of the coastline in China, including about 70% of sandy coasts, and most mudflats and coral coasts in open waters, has experienced erosion (Liu et al., 2020). Since 2014, coastline erosion in the north is less as severe as in the south (Fig. 4.12).

4.4 Governance Issues

China government has adopted a series of measures for managing and utilizing coastal habitats and living marine resources scientifically and rationally. By formulating and implementing policies, action plans, spatial planning, laws and regulations, etc., China has achieved significant progress and achievements in this capacity. China has already set up the legal system for marine conservation, which is a framework under the Constitution and built upon *the Environmental Protection Law* and *the Marine Environmental Protection Law*, with *the Fisheries Law* and *the Water Law* as the main body, and executive regulations such as *the Regulations on the Management of Land-based Pollutants' Damage to the Marine Environment*, *the Regulations on Nature Reserves* and *the Island Protection Law* as the detailed supplements. However, the existing institutional system is still not well-rounded, and there is a huge gap between

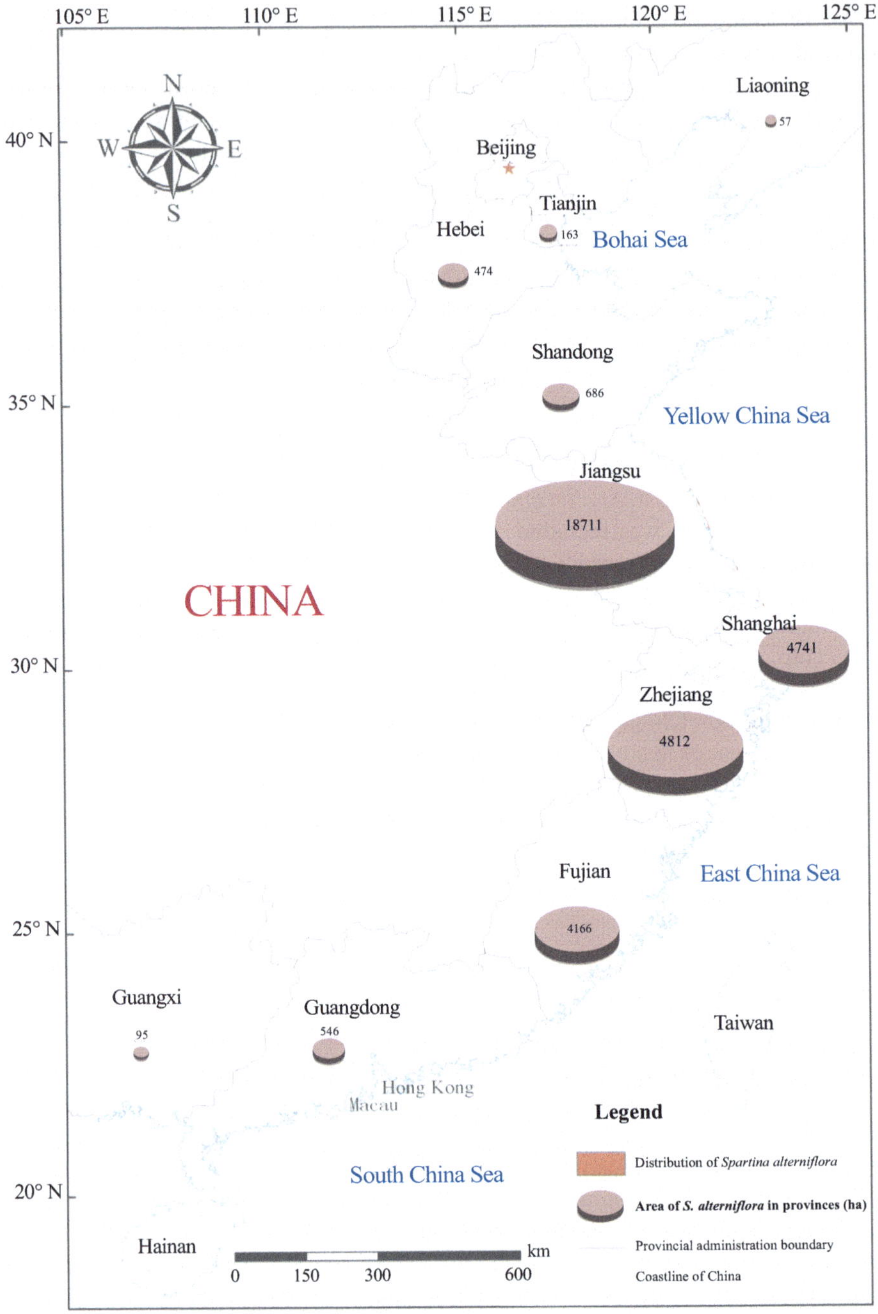

Fig. 4.11 Distribution of *S. alterniflora* along the China coast in 2007 (Zuo et al., 2012)

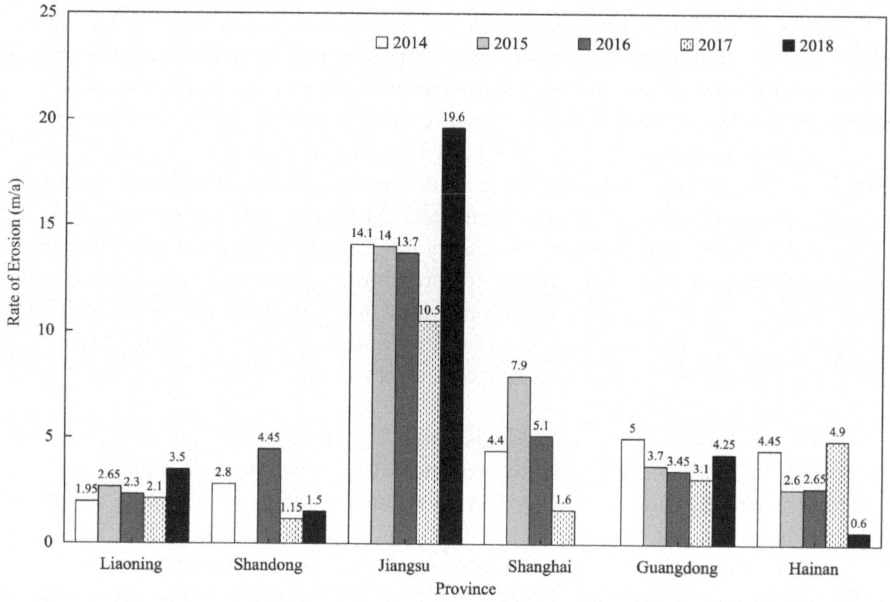

Fig. 4.12 Rates of coastal erosion in Liaoning, Shandong, Jiangsu, Shanghai, Guangdong, and Hainan provinces (*Data Source* Bulletin of China Marine Disaster 2014–2018)

the conservation targets and the results already achieved. There are existing problems hampering the conservation actions, including:

(1) Lack of comprehensive management laws for living marine resources

Firstly, it's well recognized that living marine resources have both economic and ecological value. The importance of healthy habitats and biodiversity as a basis for sustaining the renewal of living marine resources and the broader (economic, ecological and social) value of marine living resources are still not fully recognized and evaluated in China. Although the current system has played an important role in limiting the large-scale reclamation and pollution of coastal wetlands, it still has shortcomings in ecological protection. The newly amended *Marine Environmental Protection Law*, for example, has increased the penalties for pollution, with "pollution prevention and control" accounting for 5 chapters and 44 articles of the total 10 chapters and 97 articles, while "marine ecological protection" accounts for only 1 chapter and 9 articles, or less than 10% of the total number of articles. Encouragingly, the prominence given to the creation of ecological civilization in the 13th Five-Year Plan and the 2018 amendment to the national constitution set the stage for such a policy, which builds the foundation for the formulation of national legislation related to the integrated management of living marine resources, but its development has not yet commenced.

Secondly, the existing legislation lacks systematic integrity. Since China's legislation system is mainly classified according to the natural elements, it cannot organically link the ecological functions and structure of the ecosystem. In other word, the integrity of the ecosystem has not been fully respected in the legislative process. Existing relevant national policies lack key requirements related to ecosystem-based protection, for which the relationship and integrity between living marine resources and their key environmental requirements should be taken into account: (1) Identification of species and habitats necessary for maintaining ecosystem functions and values; (2) Defining requirements for protecting or restoring those species and marine ecology where possible. Therefore, based on the idea of integrated ecosystem management, it is necessary to develop a unified and coordinated comprehensive law on marine resource protection, focusing on conserving the ecological value of marine biodiversity.

Thirdly, there are still gaps in the legislation. In addition to the need for a comprehensive protection law mentioned above, specialized laws in certain areas are also needed for marine ecological protection. In regard to alien species, quite a number of aquaculture species have been introduced into China over the years, including the popular and high-value species such as *Scophthalmus maximus* and *Penaeus vanname*. None of these have been subject to robust biosafety checks, and it is unknown if any introduced aquaculture species will likely result in biological invasion. However, cordgrass *Spartina spp.*, which was introduced to China as an aquatic fodder and for shoreline protection, did become invasive in many locales. Several principle regulations about the introduction of alien species are made in *the Fisheries Law* and *the Sanitation and Anti-Epidemic Law*, while there is a lack of more operational and practicable implementation rules. Besides, there is insufficient monitoring, evaluation and control over the prevention, introduction and governance of alien species. In addition, climate change will not only affect the productivity and distribution of marine organisms (Gaines et al., 2018), but also affect the management of various industries, thereby affecting the entire management system. We should therefore fully recognize the timeliness of policies, and take these impacts into account when formulating new policies in the future.

Furthermore, marine ecological compensation system is incomplete. Although the establishment of marine reserves, as the most effective way in protecting marine biodiversity, has played an active role in the protection of typical or important ecosystems and biological species, it poses conflicts with exploitation of marine biological resources by local residents for their livelihoods and development. Besides, putting an end to the destruction of coastal and marine ecosystems and biodiversity is extremely challenging. Therefore, it is urgent to compensate for the damaged marine ecological elements and their functions, and guarding them against further deterioration and destruction. To resolve these problems, *Marine Environmental Protection Law* clearly points out that the State has the obligation to establish a compensation system for marine ecological protection. Existing common marine ecological compensation methods include: the state's subsidies for fishermen career conversion, the imposition of dumping fees and sewage charges, as well as enhancement release, fishing moratoria, and construction of artificial reefs. Although all these measures have achieved certain results, issues remain in: whether the existing financial subsidies could make up for fishermen's needs for conversion, whether the collected sewage charges could effectively prevent pollution of the marine ecological environment, and how to use and allocate the sewage charges after their collection (He & Chen, 2020), just to name a few.

(2) Inadequate supervision and management

Unclear or conflicting limits of management authority will result in ineffective or inefficient decision-making. Although new regulations related to marine habitats and biodiversity in China have been under development, they are scattered as various laws, regulations and normative documents; lacking of coordination and connectivity may result in overlaps and conflicts between these regulations (Wang et al., 2008), and these conflicts will lead to challenges in management. Besides, multiple departments involved in the protection and management of marine habitats and biodiversity often have different interest demands, and they will also manage from different perspectives in specific practices, which hinders the coordination of responsibilities among all parties, and hampers rapid and reasonable actions in protection of marine habitats and biodiversity. In recent years, China has also taken measures to deal with this problem. For example, in the construction and management of marine reserves, overlapping and multiple management is a long-standing issue, which is incompatible with the requirements of development in the new era. To solve these problems effectively, the State Council enacted *the General Plan for the Establishment of National Park System* in 2017, proposing to establish a unified authority and hierarchical management system. After the establishment of national parks, a unified management agency will be integrated and formed to perform the corresponding management responsibilities within the scope of national parks, and holistically perform resource-environment law enforcement responsibilities when necessary (as authorized by the actual situation).

In addition, public awareness and financial support are also important factors affecting coastal protection and restoration. Education and research on coastal ecosystem protection have been gradually strengthened over the past ten years, but there are still some challenges. On the one hand, public awareness is limited on the full breadth of values provided by living marine resources and the threats they are confronted with, which means civil society is not voicing its support for and contributing appropriately to conservation. On the other hand, most of the funding for protection of coastal habitats and marine biodiversity comes from central government for two main reasons: (1) local governments are faced with great pressure brought by unstable and limited sources of financial support; (2) they lack an efficient platform for investment and financing to engage the participation of social capital. Having only a single investment source will put enormous pressure on the national treasury and also fail to meet the current huge demand. Besides, the degree of openness and transparency in the use of the invested funds is also unknown. These constraints will produce a series of negative consequences including insufficient funding for scientific research and protection capabilities, and backward infrastructure, which are serious obstacles to the restoration and management of marine habitats and the conservation of biodiversity. Therefore, in the future, China should establish more efficient fiscal mechanisms, such as building long-term and diversified financing platforms, and including ecological construction funds in the budgets of the people's governments at all levels, etc. Only when there are no economic restrictions can we do better in the conservation of coastal habitats and marine biodiversity.

4.5 Technical Challenges

As the factors affecting China's coastal and marine ecosystems are multifaceted, the technical basis for comprehensive conservation of habitat and biodiversity is even more complex due to the much greater number of impacts and outcomes considered. Adding to this complexity, the value of many living marine resources cannot be quantified as readily as the revenue generated by seafood sales, making the inputs and outputs of conservation more difficult to measure. For example, a 10% increase in fishery yield won't, in most cases, correlate to an approximately 10% increase in abundance of a popular wildlife species or 10% more tourism revenue. Both the impact on and value generated by habitats and biodiversity are characterized by uncertainties, nonlinearities and complex interactions. Overcoming these challenges is hindered by insufficient research, monitoring and technical standards.

At present, insufficient surveying and monitoring is the biggest obstacle facing marine biodiversity conservation. In order to fill the existing data gaps and gradually realize the overall protection, systematic restoration and comprehensive management of China's marine habitats and biodiversity, it is necessary to build a corresponding survey and monitoring system. *The General Plan for the Construction of Natural Resources Survey and Monitoring System* issued in 2020 proposed to build a survey and monitoring standard system with natural resources classification as the core, and a technical system with remote sensing monitoring as the main approach. Natural resource classification is the basis for marine habitat and biodiversity management, as well as the premise for survey and monitoring; while remote sensing technology is advantageous in marine resources monitoring owing to its extensive coverage, rapid data update and high spatial resolution. Besides, in order to fully understand the current situation of marine habitats and marine biological resources (type, quantity, quality, spatial distribution etc.), along with their protection and utilization, a comprehensive survey on the status of China's marine resources is needed based on the monitoring system construction.

Great efforts have been invested in conducting wetland surveys in China: the first two large-scale multidisciplinary wetland resources censuses were conducted during 1995–2003 and 2009–2013. In the context of rapid economic and social development, such periodic monitoring can no longer meet the needs of wetland conservation. Therefore, on the basis of the background data of marine natural resources, we should also monitor their dynamic changes, so as to timely grasp the changes caused by human activities or other factors, and achieve the regulatory goal of "early detection, early checking and strict cracking-down" of violations. Additionally, there is no comprehensive information platform in China for the sharing of data on coastal habitat protection. The main stakeholders (wetland authorities of governments at all levels, management units for protected areas, institutes and universities engaged in coastal habitat monitoring and research, and relevant NGOs from home and abroad, etc.) lack necessary communication and cooperation, and the transparency of existing monitoring data is low; all of which hinders the development of more innovative research and the improvement of the quantity and quality of technical analysis. To solve this problem, the above-mentioned data need to be standardized and integrated, and a historical survey and monitoring database needs to be built to enable free data retrieval and exchange for scientific purposes.

Only when the data obtained from monitoring system and information sharing platforms are applied to actual research, can the protection and restoration of the coastal ecosystems be well conducted. In the last decade, great marine ecological restoration efforts had achieved certain results in China. Until 2018, 156.6 km of coastline and 35.49 km^2 of coastal wetlands had been restored, and in some typical coastal ecosystems such as mangroves, seagrass beds and coral reefs, more stable wildlife populations had developed (Liu et al., 2020). However, theoretical and technical research related to marine ecological restoration is insufficient. A large number of preliminary restoration projects have been implemented over the past decades, such as introducing species, removing invasive species, improving water quality, removing fish farms, and improving hydrological conditions. However, those efforts still lack sufficient consideration of the ecosystems in that: (1) the latest ecological principles (such as Nature-based Solutions) were not adopted; (2) the type of ecosystem problems to be resolved were not defined accurately; and (3) research on the causes of ecological degradation was not adequate. As a result, restoration measures were usually not tailored to a particular coastal ecosystem at varied degrees of degradation, so that they tend to fail to achieve the expected results. In addition, China is still in need of in depth international cooperation on coastal habitat protection and biodiversity restoration, and the introduction of advanced concepts and technologies.

Theoretical research is not enough to create a technical standard system, which is the prerequisite and technical guarantee for the protection and restoration of marine habitat and biodiversity (Liu et al., 2020). Most common marine technical standards in China are mainly for investigation and monitoring of coastal areas, and formulation of standards for water quality conditions and some environmental indicators. The former mainly includes national and industry standards such as *Specification for offshore environmental monitoring (HJ 442-2008), Index system for important wetland monitoring (GB/T 27648-2011)* and *Technical specification for coastal wetland eco-monitoring (HY/T 080-2005),* which are used to guide the routine monitoring and various thematic monitoring of coastal habitats and coastal waters. The latter mainly includes national compulsory standards such as seawater quality standards, and constraints or development planning values for certain indicators. However, because the indicator system is relatively isolated and scattered, and lacks comprehensive evaluation indicators for ecosystem health, these standards are not enough to achieve technical guidance and performance evaluation for marine ecological restoration. In 2020, the State issued 21 technical standards and specifications aimed at strengthening technical support for the protection and restoration of typical coastal habitat ecosystems such as seagrass beds and coral reefs, including 10 technical methods for coastal zone surveys and ecological assessment, 10 for ecosystem restoration and 1 for project supervision. The establishment of a standard system for marine ecological restoration not only requires researchers to formulate and improve standards in accordance with the concept of coordinated land-sea protection, systematic restoration, and comprehensive management, but also requires universal recognition and coordination among various sectors to better guide coastal protection and restoration.

Chapter 5
Climate Change as a Major Challenge to Living Marine Resources

Robert Boenish, Jacob P. Kritzer, and Fang Sun

5.1 Impact of Climate Change

Climate change has had an increasingly dominant impact on global ocean ecosystems (Halpern et al., 2008; Cheung et al., 2010; Sumaila et al., 2011; Doney et al., 2012). Changes in temperature, pH, dissolved oxygen, salinity, current patterns and other factors are reshaping patterns of growth, reproduction, survival, migration, species interactions and habitat availability in ways that are fraught with uncertainty. Although some marine ecosystems are more affected than others—in particular, coral reefs and other biotic coastal habitats that depend upon a very precise combination of environmental variables—all experience these effects to some degree. Ecosystem changes affect species that are grown through mariculture operations, harvested by fisheries or provide other living marine resource values.

Climate change affects marine organism productivity and distribution either directly, through environmental changes, or indirectly, through effects on important prey, habitats or other ecosystem components (Gaines et al., 2018). Changes in productivity affect the potential available yield for mariculture and fisheries, as well as the recovery potential of species affected by these or other factors. Changes in distribution can introduce species that might compromise fishery production and affect which species are available for local fisheries, tourism and other uses.

R. Boenish (✉)
Environmental Defense Fund, Washington, DC, USA
e-mail: robert.boenish@gmail.com

J. P. Kritzer
Environmental Defense Fund, Boston, USA
e-mail: jake@neracoos.org

F. Sun
Environmental Defense Fund, Beijing, China
e-mail: sunfang@cet.net.cn

H. Liu et al. (eds.), *Conservation and Sustainable Use of Living Marine Resources and Biodiversity*, Global Ocean Governance and Ecological Civilization,
https://doi.org/10.1007/978-981-19-4790-2_5

Climate change represents a system-scale impact that will affect management of individual sectors and more comprehensive living marine resource policies; thus, these factors should be considered to ensure that future policies remain effective as environmental conditions evolve. Owing that China's marine resources span from tropical to temperate environments, distributional shifts or changes in productivity will be particularly acute, especially near transition zones.

5.1.1 Marine Aquaculture

Much attention in recent years has been directed to the use and optimization of mariculture to help solve global food and nutrition security (Gentry et al., 2017). Indeed, food products from aquaculture have outpaced wild capture fisheries for some time, and have surpassed capture fisheries in overall food production (FAO, 2020). As with the agricultural revolution of the eighteenth and nineteenth century, aquaculture practices have undergone a rapid transition through technological breakthroughs and economic promise. Presumably, more control of the food process can translate to efficient production, market stability and improved livelihoods. This is particularly relevant because nearly 85% of marine and coastal aquaculture is produced by Asian and African countries, and China is responsible for over half the world's total (57.9%) (FAO, 2019).

In the face of climate change, nutrition and food security from marine and aquatic sources have been cited as major concerns, stemming from projected declines in fishery catches due to climate effects (Golden et al., 2016). If aquaculture could compensate and supplement nutrition in these regions, it could have substantial economic, public health, and quality of life implications. It is still unclear how aquaculture will fare in the face of changing climate, but in the past few years, researchers around the world have started tackling the question of whether aquaculture is resilient to climate change. Results so far suggest that geography, species choice, management and breeding success will all play large parts in determining success (Clavelle et al., 2019).

Analogous to siting a terrestrial farm, aquaculture siting is critical to success. Multiple attributes of water quality—e.g., temperature, concentration of pollutants, primary productivity and pH—must be naturally in the suitable range of the cultured species, or must be artificially controlled. Studies have shown different regions will have different susceptibilities to warming and ocean acidification in the coming decades (Gentry et al., 2017; Froelich et al., 2018). In China, the next few decades are projected to introduce a modest (5–15%) decrease in suitable area for finfish aquaculture outside of the central coast. For Chinese bivalve aquaculture, the negative changes are projected to be even more dramatic, particularly offshore and in the southern regions. For species like salmon, high levels of water circulation and a narrow temperature tolerance are required to mitigate disease and maintain animal health. In western countries, oyster aquaculture is emerging as one of the more progressive forms of culture. Some oyster aquaculture firms are already adapting to and anticipating changes in ocean pH. Similar to many marine organisms, oysters are more susceptible to negative effects from environmental conditions early in their life history. To combat this, some shellfish farms are artificially breeding and rearing oyster spat to get them through the critical settlement phase. Presumably, the increased survivorship is economically significant enough to warrant investment in rearing facilities.

Nearly 50% of mariculture is made up of extractive species, which are those that depend on obtaining food by filtering water. Therefore, nearly half of the world's mariculture depends on plankton productivity. Unfortunately, many studies now have corroborated that global primary productivity will likely decrease with climate change (Lavoie et al., 2010; Roxy et al., 2016). Further, changes in plankton composition can have mixed effects on nutritional input. Overall the shift has been to smaller, less lipid-rich species (Johnson et al., 2011). For example, in the Gulf of Maine, well documented regime changes in the dominant species of copepod are due to warming surface waters (Grieve et al., 2017). Researchers are already seeing these results cascade through the ecosystem. Changes in coastal productivity all the way up to changes in whale migrations and emaciation (whale depend on the lipid-rich *Calanus finmarchicus*) are now considered commonplace. Effectively, the overall primary productivity budget is the backstop for production capacity of natural systems and extractive aquaculture. For species such as oysters and clams, changes in productivity have direct causal implications for food production.

For nearly 30 years, it has been increasingly accepted by the scientific community that climate change will introduce overall increasing upwelling (Bakun, 1990; 2017). Some of the most productive marine regions exist in close proximity to zones of intense upwelling. Physically, water is able to hold less dissolved gas as it warms. And additionally, upwelling more recently has been linked with the increase of oxygen dead zones (Diaz & Rosenberg, 2008; Bakun, 2017). These two factors suggest both acute and dissipated oxygen stress will increase in coastal waters. Aquaculture may be susceptible to decreased oxygen in more profound ways than the natural ecosystem; among the most obvious is the disadvantage of operations and sites being relative immobile. Secondly, decreasing oxygen short of dead zones slows organism productivity and further changes the prey field for filtering organisms. Increasing nutrient inputs from factors such as agriculture runoff will increase the frequency of algal blooms, which themselves can cause substantial marine life die-offs. These blooms can exist in forms of both inert and harmful toxin-producing algae.

Harmful algal blooms (HABs) are much more likely to exist in the intersection of nutrient input and upwelling zones. These blooms result from the rapid proliferation of algae carrying one of a suite of harmful toxins. The toxins can cause symptoms ranging from nausea to paralysis and death (Figgatt et al., 2017). Beyond toxins, HABs can cause seawater eutrophication that can lead to massive die-offs of aquatic organisms. Literature predicts an increase in the frequency, amplitude and toxicity of HABs due to climate change (Davidson et al., 2014). Often, these occur in or near estuarine zones. Fortunately, direct threats to humans from HABs can be strongly mitigated with appropriate monitoring and environmental regulation (Chap. 4). Monitoring programs must be consistently supported and require strict red line standards to maintain food safety. Nonetheless, HABs often bring significant costs to the ecosystem and to the economy. In the United States alone, the annual economic costs of HABs is US$100 million (Davidson et al., 2014), and upwards of US$1 billion in Europe (Granéli & Turner, 2006). Economic consequences include costs associated with public health, lost product or closures, lost recreation and tourism, and management. For most parts of the world there are published or reputable studies on economic estimates, but very likely the global extent is many billions USD.

China is uniquely endowed in terms of coastlines and major river estuaries. Unfortunately, from these rivers flow acute nutrient inputs which can be directly traced to agriculture. Excessive input of nutrients such as nitrogen and phosphorous essentially act as algae fertilizer. Toxin-producing HABs are thus common place in Chinese waters including the Beibu Gulf, Yellow Sea, Changjiang River estuary, and South China Sea (Xu et al. 2019; Yu et al., 2017; Tang et al. 2006).

Presumably climate change will create winners and losers in finfish and invertebrate aquaculture. Calcifying species, for example, are vulnerable to regions projected to become more acidic. To further illustrate this case, due to increases in coastal upwelling, the US west coast is projected to decrease its bivalve aquaculture potential in most places by 2050 (Froelich et al. 2018), whereas, further north on the west coast in Canada, researchers predict an increase (Froelich et al. 2018). Additionally, many of these species depend on feeding on plankton present in the natural environment. Identifying current examples of species grappling with the pressure to calcify can help in the development of mitigation and adaptation strategies in China and abroad.

Algal and Integrated Aquaculture

Algae mariculture, or the intentional culture of seaweed, is already a multi-billion dollar industry and is poised to expand. Algal products are increasingly used in industry, fertilizer, livestock feed, and for human consumption. Recent works have commented that seaweed aquaculture is the fastest growing component of global food production (Duarte et al., 2017) and is more limited by implementation than available suitable area (Froehlich et al., 2019). Further, increased oxygen supply, fish habitat creation, and decreased acidity are suggested among a long list of positive externalities of algae mariculture.

In China, which grapples with large volumes of effluent and runoff from extensive agriculture and aquaculture waste water, algae mariculture can be a vital tool to combat coastal eutrophication. He et al. (2008) found that in Jiansu Province, culture of the rhodophyte (red algae) *Porphyra yezoensis* substantially removed excess nutrients from the water. As *P. yezoensis* is itself a multi-billion dollar industry for its use as a food crop and production of nori, bioremediation with excess farming may be not only possible, but economically advantageous. Algae mariculture may also play a future role in offsetting carbon emissions. Due to being fast growing carbon sinks, the use of kelp aquaculture is promising for fuel production, creation of non-fossil fuel fertilizers, and animal feed, all which can help offset carbon emissions (Duarte et al., 2017). It has been recently shown that in fact only a small proportion (14–25%)

of current algae aquaculture would be needed to offset carbon estimates of global finfish and crustacean aquaculture (Froehlich et al., 2019).

Integrated multi-trophic aquaculture (IMTA), which is widely practiced in China, is a possible outlet for humanity to benefit multifold. Through a variety of aquatic and marine forms, culture systems that integrate the refuge, nutrient assimilation and oxygen benefits of photosynthesizing organisms, with carnivorous species such as finfish, are gaining interest from industry. A commonly cited problem with fish pens is their inefficiency and release of nutrients. In some places, the intensity of nutrient release (e.g., nitrogen, phosphorous) is sufficient in and of itself to cause eutrophication. But by coupling this system with algal species, which can absorb some of these released nutrients, growers can contain many of the harmful crop externalities, increase efficiency and deliver a more diversified portfolio of products. However, while this technology is promising and is developing quickly, there is still much to be learned in terms of monitoring, species choice and zoning.

5.1.2 *Wild Species*

The expectation that climate change effects will differ by latitude is well documented in the literature. Fundamentally, climate change will result in regional differences. The structure and resilience of each ecosystem is similarly varied. Pinsky et al. (2013) highlighted the differences in pace of change, or "climate velocity", by region. They suggest that climate velocity is a critical factor in the ability of systems and individual species to cope. Interestingly, marine species more closely track changes in climate velocity than their terrestrial counterparts, and the ability to follow temperature clines can make some species more resilient to change (Pinsky et al., 2013). However, distributional change comes with complexity.

Change in water temperature is already affecting the patterns of oceanic currents. The degree of upwelling and mixing profoundly changes water chemistry, and ultimately, productivity capacity. As noted as early as 1912 by Johann Hjort, events in the early life history phase have the largest effects on success. For planktonic larvae to reach even a juvenile phase, there is a gauntlet of challenges that must be overcome (Arnold et al., 2010), the least controllable of which is geography. Seasons are changing, and along with them, species phenology, including changes in molt or migration timing. In the case that a species releases pelagic eggs or small, free swimming larvae, the timing influences where the eggs and larvae up. If swept too far offshore, there may be an increased chance of starvation, or the water may be sufficiently deep to ensure fatality soon after settlement. For a species to make it through the gauntlet, geography and habitat matter.

Planktonic organisms often depend on the co-occurrence of species for prey. For example, juvenile lobster depends on the availability of a wide variety of copepod species. Mismatches in timing can occur when environmental changes affect species' phenology differently. For a prey species, this is a stroke of good luck. For predators, the opposite is true. It is fair to assume that not all species will respond to the same gradients of environmental cues. Thereby, changes in the magnitude or variability of temperature may strongly affect the degree to which timing matches or mismatches occur.

While relative responses to changes between species in the United States are relatively well studied, the effects of changes in life cycle timing are still poorly understood (GAO, 2016). Much like seeing cherry blossoms in September, the biological consequences of changes in life history timing are presumably high, but difficult to characterize and predict. In the Gulf of Maine, for example, increased water temperatures have demonstrably changed the phenology of American lobster *Homarus americanus*. In recent years, warmer temperatures are resulting in a coordinated molt in the early summer. Traditionally, the molt had been more broken up across the summer months. Resulting from the early molt, many more lobsters are able to undergo a second summer molt (Staples et al., 2019). Because lobsters grow incrementally through molting, this increase in temperature effectively increases the species' growth rate. Similar things can be seen for swimming crab (*Portunus* spp.) species in China. These fast growing, valuable seafood species are fished heavily at the end of the annual Chinese summer fishing moratoria across the entire China coast. If relative settlement timing, or growth rates change as a consequence of changing temperature, it will affect the size of spawning stock, and moreover, might result in changes to the average size of crab caught and fishery operations (Lin et al., 2021).

An underappreciated factor affecting how a species will fare with a changing climate is bet hedging. This phenomenon is exhibited in nearly all species, but can be easily observed in marine environments. Choi et al. (2019) show that for intertidal organisms, exposure to direct sunlight can be hugely consequential. Based on high resolution habitat modeling, small suitable areas, or microrefugia, often exist only centimeters away from uninhabitable zones. Sometimes, the best available habitat is directly adjacent to inhospitable zones. As a response, species may occupy suboptimal habitats if they are further away from unsuitable areas. For another example, Werner et al. (2019) show that for Atlantic cod *Gadus morhua* populations in Greenland, individuals tend to occupy suboptimal habitat—even when better habitat is available nearby. They hypothesize that the immediate costs to the individual include occupying suboptimal foraging and thermal habitat. However, venturing away from a safe refuge with adequate feeding opportunities may not outweigh the potential increase in predation pressure or the risks that venturing may be fruitless. Compared to more static species, bet hedging itself may be somewhat of a luxury in response to climate change.

5.1.3 Habitats

Coral Reef Ecosystems

In light of climate change, much attention has rightly been given to studying resilience and adaptive characteristics of tropical coral reef ecosystems. Hermatypic, or reef-building, corals house the largest collection of biodiversity in the ocean. Reefs live in very extreme conditions, often exposed to intense wave action, solar radiation, heat and intense competition. Corals also depend on symbiotic relationships with photosynthetic algae in the genus Symbiodinium. When stressed, the algae are expelled, resulting in coral 'bleaching'. Prolonged exposure to heat is a known major cause of large-scale bleaching events. Due to their inability to move and reliance on broadcast spawning for colonization, tropical corals are poorly equipped to migrate to more suitable habitat as the conditions change. Further exacerbating this problem is the growing issue of sea level rise. Recent work suggests that in most regions of the world, coral growth will be outpaced by sea level rise (Perry et al., 2018). If this continues, light attenuation may be substantial enough to affect the coral's symbiont productivity, and ultimately reduce coral growth rates.

Such an unoptimistic forecast for tropical reefs will directly translate to the multitude of organisms that depend on healthy reefs for food and substrate. Indeed, without substantial mitigation of climate change and other perturbations, the loss of coral reefs will mean the loss of tremendous biodiversity. Hughes et al. (2018) suggest that rapid, large-scale changes to emissions are essential to prevent the global decline of reef systems. Most scientific literature on the topic has corroborated these findings. However, on a smaller, community scale, human interventions can themselves have profound positive effects on marine reef resilience (Steneck et al., 2019).

Maintenance of functional diversity, and thereby ecosystem resilience, in some cases can be accomplished through responsible and proactive fisheries and ecosystem management. On the Caribbean island of Bonaire, proactive banning of spearfishing, and retention of reef herbivores, has been implicated in the unprecedented ability of reef recovery following severe environmental disturbances (Steneck et al., 2019). Further, reduction of chemical and physical practices that damage the reef should be top priorities in order to mitigate effects from the climate. These harmful practices can include chemical or agricultural runoff, fishing activities and even using reefs as a ship mooring field. Monitoring in a given system is essential to understanding the key species maintaining reef function. With this information, managers can be better equipped to target effective policies that mitigate reef degradation.

Mangroves

Global mangrove systems are among the best understood, but most vulnerable, habitats. Mangroves provide a wide swath of ecosystem services, including carbon sequestration and deportation. They also act as important biodiversity hotspots for obligate organisms, as well as provide habitat for juvenile reef fish (Igulu et al., 2014). Mangrove systems are among the most productive ecosystems on the planet (Carugati et al., 2018). Humans rely on mangroves for food, timber, water quality and storm protection. For this reason, the systems have tremendous value as an extractive resource, as well as for ecosystem services. In recent times, this tension between using mangrove forests for ecosystem services and short-term extraction has favored the latter.

Mangrove deforestation continued at a rapid pace over the latter half of the twentieth century, and has only slowed in the past few decades. Among the leading causes of deforestation are the construction of pens for invertebrate aquaculture and expansion of agriculture. In Indonesia, for example, such ventures have led to a 40% decline in mangrove forests in the past three decades (Murdiyarso et al., 2015). Around the world, this trend is consistent. In China, coastal development, including land reclamation and aquaculture, have resulted in a 73% loss in mangrove forests (Paulson Institute, 2016). Beyond removal of mangroves, moderate disturbance can have profound implications for biodiversity and system function. Carugati et al. (2018) found that disturbed mangroves show diversity losses of approximately 20%, and upwards of 80% loss in microbial decomposition rates. These losses of ecosystem functions highlight the importance of protecting mangroves in the context of climate change.

Outside of direct deforestation, mangroves around the world are struggling in the face of climate change. Out of all the environmental changes, loss of sedimentation due to sea level rise is the central concern for mangrove systems (Gilman et al., 2008). Generally speaking, and most prominently for the Pacific Islands, lowering of sediment elevation coupled with a rise in sea level has the net effect of eroding mangrove systems. Moreover, coastal habitation leaves little room for landward migration. If mangroves cannot colonize land to compensate for reduced sedimentation and rising sea level, the result will be net carbon export and additional deforestation. This process has well documented downstream effects on biodiversity, system productivity and ecosystem services. In addition to reducing sea level rise, proactive mangrove reforestation and strictly limiting mangrove perturbation are key strategies to preserving the important biological and ecosystem services that mangroves provide.

Seagrass Meadows

Seagrass meadows are important systems for both biodiversity and ecosystem services. Worldwide, seagrass meadows cover nearly 180,000 km^2 and are among the most productive ecosystems on the planet (Waycott et al., 2009). China is a worldwide seagrass hotspot, home to 8765 ha of seagrass meadows representing about 30% of all known species of seagrass (Zheng et al., 2013; Liu et al., 2016). From temperate to tropical coastal regions, seagrasses are the most prominent macrophytes in most regions (Duarte et al., 2018). Seagrass not only provides key stable nursery habitats for fish and invertebrates, but also aids in nutrient cycling, erosion mitigation, carbon burial and oxygen production. Seagrass loss degrades ecological connectivity among other terrestrial and aquatic systems. Ecosystem services from seagrass meadows, from nutrient cycling alone, are valued at US$1.9 trillion per year. Environmental changes and coastal development are leading to increased rates of seagrass loss. Among stressors, temperature change is recognized as the most important factor in range limitation. Since 1990, the rate of annual loss has been estimated to be as high as 7%.

Regional and species-specific differences in climate responses are rapidly emerging. Most notably, climate change will lead to the functional extinction of a key Mediterranean seagrass species by the mid-twenty-first century (Jorda et al., 2012). In Australia, a recent marine heatwave coincided with an overall 36% loss to one of the world's largest seagrass meadows (Arias-Ortiz et al., 2018). The consequences of such losses amount to globally significant changes in sequestered carbon and marine biodiversity. Research has evaluated some ripple effects extending to coral reefs and mangrove ecosystems, but biological and nutrient connectivity to wider oceanic and terrestrial systems is likely. Due to the complexity of these systems, more research will be needed to estimate these connections. Prior to that, efforts to better monitor, protect, and restore seagrass meadows are necessary.

5.2 Adaptation Measurements of Climate Change

From the above discussion, effects of climate change will vary by species and region. In many cases, the changes will occur due to shifts in bottom-up processes, such as productivity, and the degree of temporal congruence between different parts of the food web, such as phytoplankton and zooplankton. The overall effects for a species will be dynamic and based on internal characteristics and external drivers throughout their ontogeny. More free-living and pelagic species probably possess the capacity to shift their distribution as seawater warms, but for species like coral, there may not be as many opportunities to translocate. Further, due to complex tradeoffs, simply possessing the ability to travel does not mean a species will shift. In fact, there are numerous examples that show a species' bet hedging strategy can bolster occupation of sub-optimal habitat. In many regions, these changes are unprecedented, adding to the difficulty in understanding how changes may emerge.

Predicting changes to a system outside of what has been recorded presents considerable difficulties. Even given a perfectly simple scenario where the ecosystem consisted of only one species, responses to a single perturbation, like warming, could be difficult to predict. As outlined in the previous section, organisms live in a variety of environments and must respond to a variety of stimuli, including prey/predator fields, and tradeoffs associated with every action. Therefore, if large-scale climate models predict average increases in temperature, organisms may change their behavior to balance metabolic tradeoffs, and the observed effects will be difficult to predict. The difficulty lies in the fact that most modeling approaches assume some sort of stationarity, whether that be in behavior, phenology or interspecies interactions.

Generally speaking, ecologists have documented that essentially all actions taken by an organism involve trade-offs. For example, the use of shelter may reduce the chance of an organism being preyed upon, but will also reduce its forage time. The ease or difficulty of obtaining enough food sets the bar for how much risk an organism is willing to take. Due to feeding being energy intensive, increased metabolic rates stemming from increased temperatures shift a species' energetic needs. The expected outcomes of increased energetic demand include changes in feeding frequency and selectivity of prey. For example, it might not be metabolically advantageous to consume smaller or lower calorie prey, and presumably, feeding frequency may increase. Across habitats and phyla, there exists a wide array of interesting bet hedging strategies. Choi et al. (2019) documented bet hedging for intertidal mussels *Mytilus edulis*, and found the optimal survival strategy involved occupying suboptimal habitats as a form of bet hedging. Over long periods of time, the most optimal habitat had higher predation levels, and on low tide events, would incur higher risks of thermal stress.

Most mechanistic model approaches use available species distribution datasets and assume presence is correlated with optimal habitat. These assumptions do not take into account a species' tradeoffs, or consider they may be occupying suboptimal habitats (i.e., warmer or colder than optimal). When these data are fed into climate change models, this may result in unknown biases that make predictions all the more difficult.

However, on a basic mechanistic level, trends from prediction studies are reasonable proxies for what can be expected in the future. As this branch of science is gaining a lot of attention, improved methods will certainly emerge. The importance of monitoring also suggests that future efforts will have access to more comprehensive datasets than in the past. On simply a heuristic level, looking back from only a few years in the future will give scientists new opportunities to predict from the perspective of scientists today. In China and abroad, it is therefore necessary to exhaustively study the current and ecosystem baselines and to propose creative strategies to model ecological responses to climate change. Increasingly, answers will emerge to the question: are our models robust to climate change?

Part II
Progress and the Road Ahead

Chapter 6
Progress on Policies for Management of Living Marine Resources in China

Hui Liu, Yang Han, Ling Cao, Jacob P. Kritzer, and Fang Sun

Commitment to ecological civilization has taken a firm hold in China's national policies. Through the 13th Five-Year Plan, President Xi's report delivered at the 19th National Congress of the Communist Party of China, the Constitutional Amendment in 2018 and other prominent policy instruments, China is repeatedly making a clear declaration that economic progress and social evolution will only continue in ways that ensure environmental protection. Indeed, **Achieving economic, social and environmental development in harmony, not at the expense of each other** has become the philosophy for modern China's governance. Rather, these goals can only be achieved—and, more importantly, sustained—in concert with one another. Economic progress that degrades environmental resources will undermine the basis for many economic sectors, while declining social conditions will incur costs to welfare and other wellbeing, and offset increased profits. Conversely, environmental protections that do not allow for economic prosperity will create social costs, lose public support and ultimately fail.

H. Liu (✉)
Yellow Sea Fisheries Research Institute, Chinese Academy of Fishery Sciences, Qingdao, China
e-mail: liuhui@ysfri.ac.cn

Y. Han
Development Research Center of the State Council, Beijing, China

L. Cao
Xiamen University, Xiamen, China
e-mail: caoling@xmu.edu.cn

J. P. Kritzer
Environmental Defense Fund, Boston, USA
e-mail: jake@neracoos.org

F. Sun
Environmental Defense Fund, Beijing, China
e-mail: sunfang@cet.net.cn

H. Liu et al. (eds.), *Conservation and Sustainable Use of Living Marine Resources and Biodiversity*, Global Ocean Governance and Ecological Civilization,
https://doi.org/10.1007/978-981-19-4790-2_6

The U.N. Sustainable Development Goals (SDGs) likewise aim to achieve the triple bottom line outcomes of economic, social and environmental progress. China signaled this new philosophy to the world through its clear commitment to the SDGs. In 2016, China developed and released *China's National Plan on Implementation of the 2030 Agenda for Sustainable Development*, in which it has created action plans for each SDG target. Since the implementation of the plan, China has already undergone one round of reporting on its targets. There have already been notable achievements in environmental protection, for example, in the areas of energy consumption and air pollution, which relate to SDG #13 on combating climate change. In 2016, China's energy consumption per unit of GDP fell by 5%, while carbon dioxide emissions per unit of GDP fell by 6.6% (MFA, 2017). In 2020, China pledged to reach a carbon emission peak by 2030 and achieve carbon neutrality by 2060. China has also made tremendous progress in addressing SDG #1, which is focused on poverty alleviation. Compared with the figures for 2015, poverty-stricken population in rural China had decreased from 55.75 million to 16.6 million, and per capita disposable income of rural residents in poverty-stricken areas had increased from 7,653 CNY to 10,371 CNY by the end of 2018. Although 2020 presented a dual challenge to the world with a great change not seen in a century tandem the COVID-19 pandemic, China has still managed to complete the mission of poverty alleviation in the new era as planned for in 2015 (The State Council Leading Group Office of Poverty Alleviation and Development, 2020).

With regard to SDG #14, China has taken several significant steps to conserve and sustainably use living marine resources, including drawing redlines that place 30% of China's sea areas and 35% of coastlines under protection, which effectively expanded the total area to be conserved and intensified law enforcement. China has also imposed more stringent standards for discharging pollutants into the sea, and improved pollution treatment facilities and sewage networks in coastal regions. China government has increased subsidies for reduction of fishing boats and offered subsidies for scrapping fishing vessels, by setting limits on the number and total power (the double control) of fishing boats. Finally, China is providing assistance to build aquaculture facilities in countries along the Belt and Road Initiative (BRI)—one of the country's most prominent and powerful vehicles for international engagement in the modern era—thus emphasizing the need to work towards environmental and social goals in tandem with economic goals.

In the maritime realm, the path for achieving a triple bottom line internationally through the BRI has been outlined in the *Vision for Maritime Cooperation under the Belt and Road Initiative* issued jointly by China's State Oceanic Administration and National Development and Reform Commission in 2017. The *Vision* addresses the importance of living marine resources and specifies the need to improve management of mariculture and fisheries, and conservation of habitat and biodiversity, across the region.

China's international leadership through BRI and other channels will be stronger as it develops domestic success stories that can inform policy changes abroad. To that end, a series of important domestic policy developments that are improving living marine resource management in China will play an important role. As discussed previously, even before the ecological civilization mandate was elevated during the 13th Five-Year Plan period, China began to address the unreasonable growth and environmental impacts of mariculture through new policies, such as the setting of an upper limit for total mariculture area by 2020, and the "volume reduction & value increase, quality and efficiency improvement, and green development" principles put forward by the Fisheries 13th Five-Year Plan. Permitting, monitoring and enforcement are also becoming stronger, which has lead to reduced overdevelopment in the coastal zone, although continued improvements in all of these areas are needed.

The 13th Five-Year Plan prompted the Ministry of Agriculture (MOA) to issue a sweeping and ambitious national fishery policy early in 2017. The policy set reduction targets for total catch and the double control on fishing vessels. Mechanism of fishery management at provincial and local scales were also addressed, including improved monitoring for stock assessments and ensuring compliance; use of fishing rights in management; and a shift toward output controls. Because output controls represent a fundamental change in China governance, the MOA initiated pilot projects in total allowable catch (TAC) management in five coastal provinces, and a total of 15 pilot projects were implemented as of 2020.

These new fishery policies by MARA focus not only on improved management of harvested species, but also on protection of ecosystems as a whole from adverse effects of catch fisheries. These provisions will be important steps toward a more holistic and integrated management of living marine resources. Other policies and pilot projects have also taken important steps in this direction. For example, the integrated ecosystem-based planning pilot in Xiamen has aimed to define space for mariculture, fisheries, shipping and other uses, while minimizing the impacts of these uses on one another and ensuring conservation of critical ecological resources. The pilot strives for this complex balance in a heavily urbanized coastal area, which requires a strong technical foundation and co-management by multiple agencies and levels of government. Notably, the project has been successful in protecting a small but stable population of the Chinese white dolphin, the northernmost extent of the species.

Integrated management of living marine resources might be most effectively achieved not only through holistic policies and coordinated governance, but also by harmonizing management with local traditions and values. Since 2013, China has begun moving in this direction on land through implementing the beautiful countryside concept, which formally promotes incorporation of human wellbeing, quality of life and cultural factors into policy making. Building from this concept, beautiful fishing villages can be a powerful means for incorporating local perspectives, objectives and knowledge into living marine resource management strategies tailored to the unique social, economic and ecological attributes of different places.

6.1 Existing Mariculture Laws and Regulations

6.1.1 Mariculture Management Institution

China has a relatively complete institutional system in the management of mariculture. From the 1970s to the 2000s, China has adopted positive and encouraging policies for marine and freshwater aquaculture. The *Fisheries Law of the People's Republic of China* (enacted in 1986, hereinafter referred to as *Fisheries Law*) established a fishery development policy of "focusing on aquaculture (or aquaculture priority)" (Tang, 2017). This important policy greatly promoted the rapid development of China's aquaculture industry. *Fisheries Law* is the central and essential law for fisheries administration in China, governing all aquaculture and catch fishery activities in inland waters, tidal flats, territorial waters and the exclusive economic zone of China, and in all other sea areas under China's jurisdiction. In addition, China has also promulgated and implemented a set of laws and regulations governing the development of mariculture industry and other relevant industries, including the *Law of the People's Republic of China on the Management of Sea Area Use*; *Property Law of the People's Republic of China*; *Animal Epidemic Prevention Law of the People's Republic of China*; *Marine Environmental Protection Law of the People's Republic of China*, etc.

The *Fisheries Law* was the first law governing aquaculture, making provision for the first time in China for the aquaculture permit system, with other relevant laws adopted over the following 20 years. The overarching legal framework governing aquaculture in China consists of five national laws, four administrative regulations and a number of rules and normative documents (Table 6.1). The rules and regulations are usually issued as interpretation or implementation documents for the laws. This is a hierarchical system, with the national laws having the supreme authority. National laws and regulations are translated at the provincial and municipal/county level to local administrative measures. These measures are usually more specific and stricter than the corresponding national laws and regulations. China's aquaculture legal system has played, and will continue to play, an important role in strengthening the protection, proliferation, development and rational utilization of fishery resources; developing aquaculture; protecting the legitimate rights and interests of fishery producers; and promoting the development of fisheries production (Liu, 2016).

For a long time, China has regarded aquaculture as an important industry to increase rural income and food production and promote rural employment, and has adopted a series of favorable and supportive policies to promote its development. MARA has a wide variety of supporting policies for aquaculture farmers, including subsidies for fishery fuel, financial projects for fishery resource protection and career conversion, fishery mutual insurance subsidies, subsidies for the development of aquaculture (also including subsidies for aquaculture machinery, subsidies for improved aquaculture strains or varieties, and subsidies for aquaculture bases, etc.), fishery discount loans, and preferential tax policies. These policies have

Table 6.1 China's Aquaculture Laws and Regulations (adapted from Liu (2016))

Title	Issuance authority	Enactment/ implementation date	Governing issues
Laws			
Fisheries Law of the PRC	State Council	1986, revised in 2000, 2004, 2009, 2019	Management of rights to use water/tidal flats for aquaculture
Law of the PRC on the Administration of the Use of Sea Areas	State Council	2002	Payment of sea use fee
Property Law of the PRC	State Council	2007	Establishment of aquaculture right system
Law of the PRC on Animal Epidemic Prevention	State Council	1997 Revised in 2007	Management of aquatic animal epidemic prevention
Marine Environment Protection Law of the PRC	State Council	1999	Marine environmental protection
Administrative Regulations			
Detailed Rules on the Implementation of Fisheries Law of the PRC	Ministry of Agriculture	1987	Guidelines for implementing the Fisheries Law
Animal Medicine Management Regulations	State Council	2004	Management of aquaculture medication
Administrative Regulations on Feed and Feed Additive	State Council	1999 Revised in 2011	Management of feed and feed additives
Administrative Regulations on Agriculture GMO Biosafety	State Council	2001	Management of biosafety of aquaculture genetically modified organism (GMO) products
Rules and Other Normative Documents			
Tentative Implementation Plan for Improvement of Water Area and Tidal Flat Aquaculture Permit System	Ministry of Agriculture	2002	Managing the aquaculture permit system

(continued)

Table 6.1 (continued)

Title	Issuance authority	Enactment/ implementation date	Governing issues
Opinions of Ministry of Agriculture on Stabilizing Aquaculture Rights and Advancing Aquaculture Permit Registration	Ministry of Agriculture	2010	Encouraging licensing and registration of aquaculture permits
Measures for Water Area and Tidal Flat Aquaculture Licensing and Registration	Ministry of Agriculture	2010	Aquaculture licensing and registration
Measures for the Management of Aquatic Fingerlings	Ministry of Agriculture	2005	Improving aquatic product selective breeding, and production, trade and import/export
Administrative Regulations on Aquaculture Product Quality and Safety	Ministry of Agriculture	2003	Supervising aquaculture product quality and safety
Approval Procedures for Original and Eugenic Variety of Aquatic Species	Ministry of Agriculture	1998, revised 2004	Managing production of original and eugenic variety of aquaculture species
Fisheries Water Standard of the PRC	Ministry of Environmental Protection	1989	National standard for control and prevention of fisheries water pollution and safeguarding aquaculture product quality

undoubtedly played a positive and effective role in supporting and encouraging the development of the industry.

6.1.2 Mariculture Permit System

In China, mariculture permits and sea use license are issued by the city or county level Ocean and Fisheries Bureau (now the Agriculture and Rural Affairs Bureau and Ocean Development Bureau), which are under the administration of MARA and the State Oceanic Administration (SOA; now an agency within the Ministry of Natural Resources), respectively. Due to a variety of reasons, some farms are launched without these two permits. Even when a fish farm is legally licensed, their change of cultured species or expansion of operation may not have been checked and constrained. This can easily result in overcapacity, water pollution, environmental deterioration, prevalence of diseases and increasing difficulties in guaranteeing the quality and safety of aquatic products.

In order to promote and guide proper growth of China's mariculture industry, and to minimize its various impacts and conflicts, a series of laws, administrative regulations and rules have been issued by the State Council and MARA over the last 30 years. One of the most important rules is that "all aquaculture enterprise should obtain a permit" (*Fisheries Law*, 1986). For a long time, management of aquaculture and capture fisheries in China has been license based, but without much planning in terms of space occupation, style of operation or carrying capacity. In theory, an aquaculture license can be issued wherever space is available (i.e., it is not in conflict with existing sea use activities), whether on the tidal flat, inshore or offshore. However, there have been huge numbers of fish and shellfish farms operating without any license. In recent years, quite a few aquaculture waters with high biomass of mainly fed species, especially fish farms operating at scale and positioned in areas with limited water exchange, have generated excessive pollution and other negative environmental impacts on fishery waters. These issues have raised the attention of government authorities, and helped strengthen governance of aquaculture activities and the "return aquaculture to natural tidal flats" movements. In some coastal provinces and municipalities, local governments have adopted a radical one-for-all approach in removing aquaculture facilities, where tens of square kilometers of fish farms along the coastline have been demolished.

With the development of China's economy and society, some provisions in the aquaculture laws and regulations could no longer meet the actual needs of governance, therefore a comprehensive amendment to the *Fisheries Law* is now being drafted. The forthcoming new version of the *Fisheries Law* is expected to include governance measures that support more balanced protection and development, as the conservation of aquatic genetic resources will be strengthened, aquaculture tidal flats will be better planned and protected, while conflicts over sea use shall be minimized.

6.1.3 China Marine Catch Fishery Governance

6.1.4 Input Control Measures

Input control is the main management measure of China's marine catch fisheries, consisting essentially of the license system, the dual control system, and the production and career conversion system.

The fishing license system was first established by the *Regulations on the Protection and Proliferation of Aquatic Resources*, which was enacted by the State Council of the People's Republic of China in 1979. In the same year, the State Fisheries Administration issued the *Interim Provisions on Several Issues Concerning Fishery*

Licenses. The *Fisheries Law* (1986) clearly stipulated the fishing license system. In 1989, the MOA formulated and promulgated the *Management Measures for Fishing Licenses* in accordance with the *Fisheries Law* and its implementation rules, which specified detailed regulations for fishing license management in waters under China's jurisdiction. However, China's fishing license usually regulates the power of the main engine of the fishing boat, the number of fishing gears, the type of operation, the operation area and the type of fishing, but there is no clear limit on the quantity of output. Fishermen can increase their catches by extending working hours and modifying fishery technology (Liu et al., 2007). In addition, due to historical reasons, there are still a large number of Three-No Fishing Vessels (TNFV) —ships with no number, no certificate (e.g., valid fishing vessel inspection certificate, ship registration certificate, fishing license) and no harbor registration—operating in China's coastal oceans (Han, 2018). Therefore, the system only reduced the number of fishermen to a certain extent, without necessarily reducing the total catch.

In April 1987, the *Opinions on the Control Indexes of Motorized Fishing Vessels for Offshore Fishing* was enacted by the State Council, and the "dual control" system for fishing vessels was launched. The system aims to limit the development of offshore fishing capacity and control the total output of offshore fisheries by controlling the total number and power of fishing vessels. In August 2002, the Conference on Coastal Fishermen's Production and Career Conversion, jointly convened by the MOA, the Ministry of Finance and the State Planning Commission, was a landmark of China's comprehensive implementation of the "reduction of ships and conversion of production" policy. In order to guide and support coastal provinces in the conversion of fishermen, a series of policies and regulations has been successively formulated, including the *Interim Regulations on the Retirement of Fishery Vessels*; *Interim Regulations on the Use and Management of Special Funds for the Conversion of Marine Fishermen*; and the *Regulations on the Use of Special Funds for Career Conversion of Marine Fishermen*. Approved by the State Council, the MOA issued the target for controlling the intensity of marine fishing during the period of the 8th Five-Year Plan, 9th Five-Year Plan, 2003–2010, 12th Five-Year Plan and 13th Five-Year Plan in 1992, 1996, 2003, 2011 and 2017, respectively (Han, 2018).

However, for more than 30 years since its implementation, the dual control system has not achieved the target of reducing the number and power of fishing boats; on the contrary, both have been on the rise. In 2017, the number of motorized marine fishing boats in domestic waters exceeded 160,000, with a total power of 11.23 million kw, showing an increase of approximately 26% and 226% from 1987, respectively (BOF, 2018). The total fishery labor force had not significantly decreased, and the decline of China's marine fishery resources had not significantly improved. Overfishing and fishery water pollution were still a normality, which was far from the goals designed by the original policy (Zhu et al., 2009; Han, 2018).

Because of this, MOA announced the *Notice on Further Strengthening the Management and Control of Domestic Fishing Vessels and Implementing Total Biomass Control over Marine Fishery Resources* in 2017, further stated that by 2020, a total of 20,000 motorized marine fishing vessels with a power of 1.5 million kilowatts (based on the 2015 control number) would be reduced. The annual reduction in the coastal provinces would be minimally 10% of the total reduction targets of the province. The document also put forward the goal of China's total marine fishing output control: by 2020, total domestic marine fishing production would be reduced to less than 10 MT, and the reduction in coastal provinces would be at least 23.6% of 2015. In principle, the annual reduction shall be more than 5%, and a provincial control index for marine fishing yield was given at the same time.

6.1.5 *Output Control Measures*

Output control is a management measure that directly regulates the amount of resource development, by regulating the total output of resources, such as the total catch of marine fisheries. In China, output control mainly includes the total allowable catch (TAC) system and the quota control system.

In 1999, China's target for zero increase in marine fishing output was regarded as the beginning of a quota system. However, it was not yet a system, but rather a goal for management. The revised *Fisheries Law* (2000) formally proposed the implementation of a fishing quota system in China's marine fisheries, but there has been no quantitative policy goal in the management practices. The 2004 *Decision of the Standing Committee of the National People's Congress on Amending the Fisheries Law of the People's Republic of China*, on the basis of continuing to emphasize the fishing license system, clearly stated that "in accordance with the principle that the catch should be lower than the increase in fishery resources, the total allowable catch of fishery resources is subject to a catch quota system." In 2016, the MOA issued the *13th Five-Year Plan for the Development of National Fisheries (2016–2020)*, which clearly stated that the total amount of marine capture fishery should be controlled at 10 MT by 2020. This meant that the 13.14 MT in 2015 should be reduced by 3.14 MT within five years, and all coastal regions were required to reduce the number of fishing vessels and control fishing intensity in accordance with the dual control and production conversion systems.

After the trial and exploration in the first two years of the 13th Five-Year Plan period, China's total marine fisheries catch control achieved certain results. In 2017, the total national marine catch fell 5% from the previous year. However, due to a long-standing lack of continuous scientific survey and monitoring, evaluation and accurate assessment of marine fishery resources, China's total marine fishery resources control system can only allocate fishing quotas based on the historical catch data of each province's statistical data. This quota allocation method is not scientific and does not completely match the actual allowable catch of marine fishery resources, and thus it is difficult to manage and regulate through administrative means (Han, 2018).

At the beginning of 2017, the Ministry of Agriculture successively issued the *Notice on Further Strengthening the Management and Control of Domestic Fishing Vessels and Implementing Total Biomass Control over Marine Fishery Resources*, and the *Letter on the Implementation of Pilot Quota Control for Marine Fishery*

Resources, and pilot fishing quota management by marine species were launched. Shandong and Zhejiang Provinces took the lead, using jellyfish in Laizhou Bay and swimming crab in northern Zhejiang fishing ground as the pilot quota control fishing species. In 2018, the pilot trial was extended to Liaoning, Fujian, Guangdong and other provinces. At present, eleven coastal provinces have independently and successively launched TAC pilots, by selecting exemplary local marine fishing species.

However, difficulties arise due to the wide variety of fishery resources, fish migration and other complex factors in the China seas. There is generally a lack of continuous monitoring data, scientific survey and evaluation of specific species and marine fishery resources in the traditional governance measures, rendering the current quota control system without science-based and reasonable principles and methods for quota allocation. Therefore, this system has not yet been fully promoted, and its effectiveness is still being explored in the pilot areas (Han, 2018). In addition, since the pilot quota control system marks the beginning of a "systematic institutional change" in China's catch fishery management, it is necessary to adopt appropriate performance evaluation to check the results and to bring out the experience of the pilot trial in a timely manner. Fishery governance is a complex systematic project, and the establishment of a scientific evaluation system is of great significance for improving both the fisheries quota control system and fishery governance capabilities.

6.1.6 Technical Control Measures

So far, input control and technical control are the earliest, and still the most widely used, fishery management measures in the world. The technical control system in China mainly includes the summer moratorium system and the management of fishing gears and methods.

Of all China's fishery management approaches, the summer moratorium (or seasonal fishing ban), has the widest spatial coverage, greatest impact, largest involvement of fishing boats and fishermen, heaviest management tasks and the most substantial protection and management measures. In order to prevent the further deterioration of living marine resources and ultimately achieve the goal of recovering fishery stocks, China began to implement this system in the East and Yellow Seas in 1995, with a subsequent extension to the South China Sea in 1999. As of today, the summer moratorium system has been fully implemented in all of China seas. From 1995 to 2018, the system was adjusted and improved 13 times; for example, in 2017, the moratorium period was extended to 4–4.5 months. However, after the moratorium, fishermen often increase their fishing intensity and reach a fishing peak, which in turn offsets the effect of the moratorium. The moratorium system provides a limited time for growth and reproduction of fishery resources, and cannot fully achieve the goal of fishing effort control (Hou & Feng, 2010; Huang & Tang, 2019).

In order to strengthen the management of fishing gear and fishing methods, in protection of fishery resources, the MOA implemented the *Notice on the Implementation of the Minimum Mesh Size System for Marine Fishing Gear* in 2004. To further alleviate pressure on fishery resources, in December 2013, the MOA issued the *Notice on the Implementation of the Minimum Mesh Size System for Marine Fishing Gear and Transitional Fishing Gear*, which proposed guidelines for the minimum mesh size of fishing gear and transitional fishing gear for marine fishing in the four China seas. Restricting the catch of juvenile fish is one of the most basic measures for conservation and management of fishery resources, but in practice, the minimum mesh size, catchable limits and juvenile fish ratio management have not been effectively implemented in China. For example, it is required that the minimum mesh size of the bottom trawl net bag in the East China Sea be larger than 54 mm, but in practice, the net size of the double bottom trawl net bag is generally 26–32 mm, and the banned double layer bag nets are commonly used. As a result, the one-year-old hairtail juveniles may sometimes comprise more than 70% of the actual catch (Zhang et al., 2007; Tang & Zou, 2010). The Yellow Sea Fisheries Research Institute, Chinese Academy of Fishery Sciences, with the support of related projects, completed an investigation of offshore fishing gear and fishing methods, improved the database of fishing gear and fishing methods, and conducted selective tests on different fishing gear materials and mesh sizes in offshore gillnets, thereby revealing the severe harmful effects of over-limit fishing gears.

Marine capture fisheries in China are characterized by several factors: a large number of fishing vessels and fishermen, many of whom have lower literacy, technical levels and awareness of policies and regulations; high cost of fishermen's conversion to other lines of business; operation across widely dispersed fishing grounds; frequent maritime transactions of fisheries products; and strong by-catch accompanied by the diversification of fishing species (Yang, 2005). All of these render China's fishery management much more complicated than that of any other country in the world (Huang & Tang, 2019). In particular, TAC management faces multiple challenges. There are currently more than six million fishermen and nearly 200,000 marine fishing vessels in China, operating in a vast water body spanning from inshore to offshore and the continental shelf. The operation methods of these fishing vessels mainly include trawling, seine nets, gillnets, trap nets, angling and others, targeting nearly 200 benthic and pelagic species of fish, crustaceans and shellfish. Compared to the fisheries in many other countries, which are relatively small in scale, simple in fishing gear and method, and mostly fishing on a single species, managing marine fisheries in China is extremely difficult.

6.1.7 Living Marine Resource Conservation in China

6.1.8 Enhancement Release of Fisheries Stocks

The *Outline of the Action Plan for Conservation of Aquatic Living Resources in China* (OAPCALR, 2006) recognizes that the proliferation of fishery resources is an important component of the conservation of aquatic living resources. In order to slow down and reverse the trend of severe decline in fishery resources, during the 12th Five-Year Plan period, the *National Fisheries Ecological Restoration Project Construction Plan* was compiled, which duly increased the support for proliferation and conservation of fishery resources. For example, the MOA (now MARA) has successively launched more than 10 Industry Specific Research Programs related to enhancement and conservation of fishery resources. The Ministry of Science and Technology has also launched a number of projects, such as the National Science and Technology Support Program and International Cooperation Project, focusing on research topics covering key technologies and common technical problems surrounding resource proliferation and conservation, and habitat restorations in the coastal oceans, freshwater and inland lakes. In 2013, the State Council convened a national video conference on the Modern Fishery Construction Initiative, designating five major industrial systems as major components of modern fisheries, including aquaculture, capture fisheries, aquatic products processing and circulation, enhancement fisheries and recreational fisheries. The promotion of enhancement fisheries and recreational fisheries will, in fact, provide solution to alleviate the negative impact of catch fisheries on living resources.

In 2016, the MOA issued the *Guiding Opinions on Improving Aquatic Enhancement Release during the 13th Five-Year Period*. The document stated that aquatic organisms should be proliferated through scientific measures, such as general and integrated planning, rational execution, scientific evaluation, strengthened supervision and extensive publicity. The intent was for a scientific, standardized and orderly development of the enhancement release to be realized; effective restoration and sustainable use of living aquatic resources to be accelerated; and the construction of water body ecological civilization and the sustainable development of modern fisheries to be promoted. The *Guiding Opinions* pointed out that by 2020, a complete system of aquatic living resource release enhancement would be well developed, with "distinct regional characteristics, clear targets, scientific and rational implementation, effective evaluation system, efficient and standardized management, and significant comprehensive benefits". In the past two to three decades, the number and species of marine fish released in China have increased significantly. During the "Tenth Five-Year Plan" period, China released a total of about 17.9 billion marine fish; while during the "Eleventh Five-Year Plan" period, the number of released marine fish was about 36 billion in total, doubling the number of the 10th FYP period. During the 12th Five-Year Plan period, China invested a total of nearly five billion CNY in proliferation, and released more than 160 billion units of aquatic organisms. The number of species of aquatic organisms that have been released into China coastal and freshwater bodies has also been increasing over time. In 2000, there were less than 20 aquatic species released for enhancement all over the country. In 2013, there were 245 species of (marine and freshwater) aquatic organisms released, including 199 economic species (of which 138 were fish) and 46 endangered species. The 13th Five-Year Plan identified 230 aquatic species suitable for enhancement release all over the country, including 64 species of rare and endangered species. In addition, China is currently promoting the internationalization of enhancement release of aquatic organisms through the Belt and Road Initiative, and has already launched China-South Korea and China-Vietnam joint enhancement release campaigns.

In recent years, in support of the release enhancement of living marine resources, Chinese researchers have selected key species for resource enhancement in the Yellow Sea, East China Sea and South China Sea; established ecological carrying capacity models of these key species in natural seawaters and at ecological controlled (enhancement release) areas; and evaluated their proliferation capacity in different sea areas. Rapid genetic detection methods have been developed for key resource enhancement species, and the genetic risk assessment framework for enhancement and release was constructed. At the same time, the appropriate marker and release-recapture technology for different proliferated stocks were developed. The proliferation effect of different key species by release enhancement was evaluated, and the economic benefits were estimated according to the proliferation effect and the input–output ratio. At present, the main marine species for enhancement release in China include Chinese shrimp, jellyfish, scallops and a number of marine finfish. During the National Fish Releasing Day campaign in 2018, more than 300 simultaneous enhancement release activities were held nationwide, and nearly five billion aquatic species were released.

6.1.9 Artificial Reefs, Sea Ranching and Recreational Fishery

Lately, China has attached great importance to the construction of coastal sea ranches and carried out a number of major research projects in relation to sea ranching. Sea ranches are artificial marine (mini-) ecosystems that serve several functions, including environmental protection, resource conservation and sustainable fishery output. They are constructed in suitable sea areas based on ecological principles, and use modern engineering technology and management models to make full use of natural productivity, through habitat restoration and artificial proliferation (Yang et al., 2019). China's sea ranching construction mainly includes three activities: marine ecological restoration, enhancement release, and construction of artificial reefs. Artificial reefs are simulations of natural habitats by engineering approaches, designed to protect, proliferate or rehabilitate components of the marine ecosystems (Tang, 2019). Development of sea ranching in China has progressed through three main phases, beginning with artificial reef construction, then adding enhancement release, and finally forming sea ranching (Yang, 2016). By 2018, 86 state-level demonstration sea ranches had been built nationwide, and most of them introduced a management mode of artificial reef plus enhancement release. In 2017, the *No. 1 Document of the Central Committee* (that is, the first official document released at the beginning of a year, indicating its importance, which usually addresses agriculture and rural issues) proposed the development of modern sea ranches to the whole nation. In 2018, the No. 1 Document further highlighted "building modern sea ranches"; in 2019, the No. 1 Document again emphasized "promoting the construction of sea ranches". All of these policies indicate that China government has attached great importance to the development of modern sea ranching.

In many coastal areas of China, sea ranching has become a new growth point for the ocean economy in recent years; it is an important support for the integration of primary, secondary and tertiary industries, as well as an important approach for coastal communities to conserve living marine resources, restore marine ecosystems and realize the transformation and upgrading of the marine fisheries industry. According to incomplete statistics, up to 2016, 4.45 billion CNY had been invested in the construction of 148 sea ranches in the Yellow Sea and Bohai Sea regions. These sea ranches cover a total sea area of 346.7 km^2, with 18.0 MCM (million cubic meters) of rocks or cement blocks dropped onto 157.1 km^2 of seabed, creating artificial reefs (MARA, 2017). The reefs vary from simple to more complex models, such as rare living marine resource enhancement artificial reefs, fish conservation reefs, seaweed reefs, seagrass bed and compound reefs integrating habitats for abalone, sea cucumbers, sea urchins, mollusks, fish and recreational fisheries. The compound reef is characterized by functions such as material circulation (multiple trophic level-integrating enhancement and development, with rare living marine resource as the main output), and supports recreational fishery functions. Sea ranches in these regions are of proliferation and recreational types.

Over the same period of time, 383 million CNY had been invested in the East China Sea region for the construction of sea ranches. Twenty-three sea ranches had been built, covering a sea area of 235.7 km^2; 700,000 m^3 of rocks and blocks were dropped onto an area of 206.2 km^2 of seabed to create artificial reefs (MARA, 2017). The main types of reefs in this region include: three-dimensional compound reefs, seaweed/seagrass beds and coastal island reef fish, crustaceans and recreational fisheries. Sea ranches in this region are mainly of conservation and recreational types.

Similarly, 745 million CNY had been invested in the South China Sea region for the construction of sea ranches. Seventy-four sea ranches were built, involving 270.2 km^2 of sea area; 42.191 MCM of artificial reefs were set up, covering an area of 256.6 km^2 on the seabed (MARA, 2017). The type of sea ranching in this region integrates ecological artificial reefs, seaweed beds and a compound reef with combinations of commercial mollusks, tropical and subtropical high-quality fish and recreational tourism. Sea ranches in this region are mainly for conservation and ecosystem protection, as well as production of fish, crustaceans and mollusks, with recreation and sightseeing as secondary functions.

It is estimated that the established sea ranches can generate 31.9 billion CNY in direct economic benefits, and 60.4 billion CNY in ecological benefits each year, with an annual carbon sequestration of approximately 190,000 tons, and a reduction of 16,844 tons of nitrogen and 1,684 tons of phosphorus. In addition, according to statistics, through the combination of sea ranches, coastal tourism and sightseeing, and recreational sea angling, more than 16 million tourists can be accommodated annually. It is also a common practice along China's coastline to combine recreational fisheries and the construction of beautiful villages with sea ranching. Due to this reason, the scale of recreational fisheries across the country has continued to expand, and the total output value of recreational fisheries has maintained an annual growth rate close to 20% for several consecutive years (BOF, 2020; NCSFTP, 2020).

As one of the main bodies of ocean economy, Large Gardens and Beautiful Villages construction has been promoted in Zhejiang Province in 2017; following the overall guideline of "adapting measures to local conditions, rational planning, forming features, and exemplary demonstration", a number of provincial and national-level recreational fishery demonstration bases (fishery parks) and Beautiful Fishing Villages were built. Recreational fisheries have become the new development point for transformation and upgrading of the fisheries industry, increases in fishery efficiency, and fishermen conversion. In 2017, the total output of recreational fisheries in Zhejiang Province was 2.4 billion CNY. The area accommodated 12.3 million tourists, and comprised 2,406 recreational fishery business entities with more than 20,000 employees. The province also has 786 recreational fishing boats, a recreation base area of 18.7 thousand ha, 467 cultural scenic spots and 2,269 professional reef (boat) fishing programs. At present, the province has 202 national and provincial-level, well-known brands of recreational fishery; among them are 28 national-level *Most Beautiful Fishing Villages*, recreational fishery demonstration (exquisite) bases and influential festivals, and 174 provincial-level recreational fishery demonstration (exquisite) bases (Zhejiang Ocean and Fishery Bureau, 2018).

6.1.10 China's Marine Protected Area (MPA) and Marine Aquatic Genetic Resource Reserves (MAGRR)

Biodiversity is fundamental to the health of marine ecosystems, and is an important material basis for the development of marine fisheries; establishing marine protected areas (MPAs) is one effective way to conserve biodiversity (Edgar et al., 2014). MPAs are representative natural zones that a state or social organization delineates for special protection of natural marine resources and ecological environment. They are one of the key measures to protect marine biodiversity and prevent the deterioration of marine ecosystems (Zeng, 2019). As an ecosystem-based approach to resource management, MPAs are of interest to international organizations and countries, such as the Food and Agriculture Organization (FAO) of the United Nations, the European Union, Australia and the United States. For decades, MPAs have been promoted as an aid to fisheries management. At present, there are three main types of MPAs in China, including Marine Nature Reserves (MNR), Special Marine Reserves (SMR) and Marine Aquatic Genetic Resource Reserves (MAGRR).

- Marine Nature Reserves (MNR)

The establishment of MPAs in China can be traced back to the Snake Island Nature Reserve, delineated in the Bohai Sea in 1963 (Zeng, 2019). In 1980, with the approval of the State Council, this Marine Nature Reserve (MNR) was upgraded to Snake Island-Laotieshan National Nature Reserve, in conjunction with Laotieshan Island. This event symbolizes the launch of construction and management of China's MPAs, and since then, the number of state-level MNRs has slowly grown over time. In 1995, China formulated the *Administrative Measures for Marine Nature Reserves*, to implement the principle of conservation-oriented, moderate exploitation, and sustainable development for these protected areas. For the construction and management of marine nature reserves, MNR of all types were spatially divided into core areas, buffer zones, and test areas. In 2018, MPAs were unified into the nature reserve construction system with national parks as the main body. By the end of 2018, China had established 271 MNRs at all levels, with a total area of 124,000 km^2, including 106 national-level MPAs. The total protected areas cover 4.1% of the marine areas under China's jurisdiction (MEE, 2020).

- Special Marine Reserves (SMR)

In 2005, the State Oceanic Administration approved the establishment of the first national-level marine special protection area, Ximen Island National Marine Special Protection Zone in Yueqing City, Zhejiang Province. This signified the beginning of the Special Marine Reserves (SMR) system. According to the *Interim Measures for the Management of Special Marine Protected Areas* issued by the State Oceanic Administration in 2006, a SMR refers to "areas with special geographic conditions, ecosystems, biological and non-biological resources, and special needs for marine development and utilization. Therefore, protective measures and scientific development methods shall be used for their special management." According to relevant regulations, areas that can be established as SMR may include: 1) sensitive and fragile marine ecosystems and areas with important ecological service functions; 2) areas where resource density is high or the type is complex (i.e., several marine industries are involved and development intensity is high, so that coordinated management is required; 3) areas where marine resources and ecosystems are in urgent need of restoration and improvement. Since the establishment of the first national Special Marine Reserves (SMR) in 2005, the number of state-level MPAs has increased rapidly.

- Marine Aquatic Genetic Resource Reserves (MAGRR)

China's Aquatic Genetic Resources Reserves (AGRR) refers to the defined water bodies or tidal flats, along with the necessary land area, for special protection and management of the spawning grounds, nursery grounds, feeding grounds, wintering fields and migratory passages of a particular aquatic species. The purpose of setting up

AGRRs is to protect and rationally use the aquatic genetic resources and their habitats. China's AGRRs are divided into, and subject to, national and provincial levels of administration, each with its own standards of management and conservation. The national AGRR is usually designated to an aquatic species with very important status at home and abroad, which has important economic value, genetic breeding value or special ecological protection and scientific research value. All AGRR are approved and announced by the MARA. The review experts for AGRR consist of experts in fisheries, environmental protection, water conservancy, transportation, marine and biological protection, so as to reflect the interdisciplinary and integrated quality of AGRR. So far, MARA has announced eight batches and a total of 458 national-level AGRRs, 52 of which are Marine Aquatic Genetic Resource Reserves (MAGRR).

At present, there is still a gap between China's marine and terrestrial protected areas, in terms of management mechanism, research and advancement. Compared with the well-developed national park and nature reserve systems in countries such as the United States, Canada, Australia and the United Kingdom, the management of China's MPAs is less rigorous, and the development status is inconsistent among different regions and administrative hierarchies. Generally speaking, several key aspects for construction of MPAs in China still need improvement, including site selection and spatial planning, protective measures and conservation ideology, and supervision and law enforcement. To improve the efficacy of conservation of living marine resources and their key habitats, the planning and management of MPAs in China should fully integrate geography, oceanography, ecology and other disciplines. There is an urgent need for in-depth research on the evaluation of the status of MPAs, as well as the outcomes of their protection and management.

6.1.11 New Policies and Trends in Management of Living Marine Resources

6.1.12 Eco-Conservation

Commitment to ecological civilization has taken a firm hold in China's national policies. The philosophy underlying modern state governance policies in China recognizes that achieving economic, social and environmental outcomes does not occur at the expense of one another. Rather, these goals can only be achieved, and more importantly sustained, in concert with one another. China signaled this new philosophy to the world through its clear commitment to the U.N. Sustainable Development Goals (SDGs), which likewise aim to achieve the triple bottom line outcomes of economic, social and environmental progress. With regard to SDG #14, China has taken several significant steps to conserve and sustainably use the living marine resources, including drawing redlines for conservation that place 30% of China's sea areas and 35% of coastlines under redline protection. It has expanded the total area under protection and intensified law enforcement; imposed more stringent standards for discharging pollutants into the sea; and improved pollution treatment facilities and sewage networks in coastal regions.

In the maritime realm, the path for achieving a triple bottom line internationally through the BRI has been outlined in the *Vision for Maritime Cooperation under the Belt and Road Initiative* issued jointly by China's State Oceanic Administration and National Development and Reform Commission in 2017. The *Vision* addresses the importance of living marine resources and specifies the need to improve management of mariculture and fisheries, and conservation of habitat and biodiversity across the region.

China may contribute to the sustainable development of living marine resources globally through BRI and other channels, if China can achieve significant progress in conservation of living marine resources in its inshore and offshore waters, and share this experience with BRI countries. To that end, a series of important domestic policy developments that are improving living marine resource management in China will play an important role. As discussed previously, even before the ecological civilization mandate was elevated during the 13th Five-Year Plan period, China began to address the unreasonable growth and environmental impacts of mariculture through new policies, such as the setting of an upper limit for total mariculture area by 2020, and the "volume reduction & value increase, quality and efficiency improvement, and green development" principles put forward by the Fisheries 13th Five-Year Plan. Permitting, monitoring and enforcement are also becoming stronger, which is leading to reduced overdevelopment in the coastal zone, although continued improvements in all of these areas are needed.

6.1.13 Marine Aquaculture

In January 2017, the *13th Five-Year Plan for the Development of China's Fisheries* was issued, proposing the development goals of "improving quality and efficiency, reducing volume and increasing income", as well as to "improve the planning for aquaculture waters and tidal flats" and "scientifically conduct aquaculture zoning and define the restricted and prohibited aquaculture zones, rationally arrange marine aquaculture layout, adjust and optimize freshwater aquaculture, stabilize the basic aquaculture waters, and scientifically determine the aquaculture carrying capacity and cultured species". With the implementation of relevant policies, many coastal areas across the country have carried out the goals of returning fish farm to tidal flats and coastal zone remediation, as many fish farms have been demolished. This might have been the cause of a significant decline in the annual growth rate of China's mariculture production (down from an annual increase of around 4–6% in the previous years to less than 2% in the last three years). According to FAO's (2020) forecasts, China's share of global fish farming production will further decline in the next few years, from 57.9% in 2018.

Since 2018, coastal cities and counties across the country have carefully carried out the zoning for mariculture waters —that is, the spatial planning for aquaculture permitted, restricted and prohibited areas—and have successively compiled the local mariculture water and tidal flat spatial plan (2018–2030). However, after zoning for the restricted and prohibited mariculture areas, the space available for mariculture will face a major reduction, some extensive mariculture methods will be constrained, and the development of fisheries communities and the increase of fishermen's income may face greater challenges. At the same time, some traditional mariculture waters, and fish farms with or without aquaculture permit and sea use licenses, are facing imminent pressure for conversion of production or even change of businesses. The policy transition from "aquaculture priority" to "returning fish farm to tidal flats", has posed a series of challenges to government authorities and aquaculture operators. In order to relieve pressure and boost the confidence of the stakeholders, in the *Opinions on Accelerating the Promotion of Green Development of Aquaculture Industry* issued in 2019, the MARA once again emphasized these pertinent principles for development of the aquaculture industry: carry out the concept of green development throughout the whole process of aquaculture production; implement an ecologically sound and healthy aquaculture system; give full play to the ecological service functions of aquaculture in the management of landscape, forest, field, lake and grassland ecosystems; and vigorously develop high-quality, featured, green and ecologically sound aquatic products.

6.1.14 Marine Capture Fisheries

The fishery fuel price subsidy policy was implemented since 2006, with a wide coverage, large scale and long duration. This policy has distorted the price signal and contradicts the goal of fishermen's conversion and reduction of ships. Therefore, in 2015, the MOF and the MOA jointly issued the *Notice on Adjusting Domestic Fishery and Aquaculture Fuel Price Subsidy Policies to Promote the Sustainable and Healthy Development of Fisheries.* This policy stipulated that the subsidy method should be adjusted, and the subsidy for reducing and scrapping fishing vessels should be increased; an annual implementation plan was made for a general reduction in marine fishing boats and power by about 10%, as well as a goal and a yearly plan for the renewal and transformation of fishing boats. Meanwhile, the policy also proposed to increase the crackdown on TNFVs and various types of illegal fishing operations, and to address non-compliance with ship certificates, illegal use of fishing gear and pollution of fishery waters.

The 13th Five-Year Plan prompted MOA to issue a sweeping and ambitious new national fishery policy early in 2017. The policy sets national targets for reductions in total catch and the number and total engine power of fishing vessels. Furthermore, the policy attends to the detailed mechanisms of fishery management at provincial and local scales, including improved stock assessments, monitoring to support stock assessments and ensure compliance with regulations, use of fishing rights and opportunities for collaboration in management by the fishing industry, and a shift toward greater use of output controls. Because policy evolution in China relies so strongly on pilot projects that enable provincial and municipal governments to figure out how to operationalize national objectives, and because output controls represent such a fundamental change from how China has managed its fisheries to date, the MOA policy also initiated pilot projects in total allowable catch (TAC) management in two coastal provinces at the beginning, including Zhejiang and Shandong, then extended to all coastal provinces and municipalities, by showcasing with a range of different fishery species.

Chapter 7
International Experience in Management of Living Marine Resources

Jeff Young, Hui Liu, Jacob P. Kritzer, and Fang Sun

According to the 2020 State of World Fisheries and Aquaculture (FAO, 2020), global fish production reached a record high of 178.5 MT in 2018, with marine capture fisheries accounting for a record high of 84.4 MT and mariculture contributing 30.8 MT. However, when considering the health of global marine fisheries and their ability to continue providing food, income and livelihood for coastal communities worldwide, the picture is not so optimistic. Around 40% of the world's assessed fisheries are fully exploited, meaning that target yields are being obtained but no additional yield is likely, while another 40% of assessed fisheries are overexploited or collapsed, which means that current yields are either unsustainable or have already declined (Costello et al., 2016). The proportion of overexploited and collapsed fisheries might actually be higher given that many countries do not have the capacity to assess their fisheries, and unassessed fisheries, which are not included among the global trends, are more likely to be depleted (Costello et al., 2012). Without improvements in management, such impressive production numbers are unlikely to be sustained into the future (Fig. 7.1).

J. Young (✉)
Environmental Defense Fund, San Francisco, USA
e-mail: jyoung@edf.org

H. Liu
Yellow Sea Fisheries Research Institute, Chinese Academy of Fishery Sciences, Qingdao, China
e-mail: liuhui@ysfri.ac.cn

J. P. Kritzer
Environmental Defense Fund, Boston, USA
e-mail: jake@neracoos.org

F. Sun
Environmental Defense Fund, Beijing, China
e-mail: sunfang@cet.net.cn

H. Liu et al. (eds.), *Conservation and Sustainable Use of Living Marine Resources and Biodiversity*, Global Ocean Governance and Ecological Civilization,
https://doi.org/10.1007/978-981-19-4790-2_7

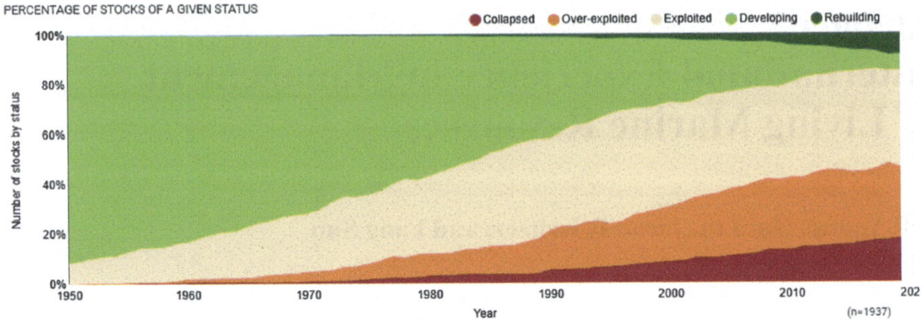

Fig. 7.1 Status of World Fish Stocks, 1950–2015 (*Source* Pauly and Zeller, 2015)

Because fishing fleets are vast and widespread, and cause direct mortality to harvested organisms—often with incidental impacts on habitat—overfishing remains the major threat to ocean health and biodiversity at the global scale. One important driver of overfishing is that many countries, especially in the developing world, lack the scientific expertise, governance capacity and financial resources needed to manage their resources sustainably. This has often led to what is known as overcapacity, or excess amount of effort and energy being applied to extracting fish, relative to the amount of fish that are available. Each year, fishermen are taking out more fish than the population can sustain, and switching to different species whenever certain species become depleted, in a pattern known as "fishing down the food web" (Pauly et al., 1998; Pauly & Palomares, 2005). Of course, other stressors also affect fishery resources, including pollution, coastal development and, perhaps most significantly, climate change.

Indeed, recent estimates suggest that ending overfishing can promote recovery of marine wildlife, including mammals, birds and turtles, illustrating the severe ecosystem-level effects of overfishing (Burgess et al., 2018). In addition, those fisheries that have collapsed represent a potential source of additional yield, if stocks can be rebuilt and renewed harvests are at sustainable levels. For example, bioeconomic modeling suggests that improved fishery management can counterbalance impacts of climate change, which, depending upon the severity of climate change the world experiences, could enable overall worldwide fishery yields to be maintained, or perhaps even modestly increased (Gaines et al., 2018). In addition, by addressing pollution and urban runoff, and mitigating or reversing the impacts of coastal development, the health of habitats that support fish populations can improve.

While the struggles of overfishing and mariculture management have been globally shared experiences, the solutions and technical expertise accumulated by developed countries in addressing those issues have not been as universal. In fact, research suggests that as developed countries began to regulate their fisheries throughout the 1990s, there was a shift in focus to the largely unassessed fisheries of developing countries (Worm et al., 2009). Developing countries' fish production, both wild caught and farmed, has doubled in the last 30 years and now accounts for over half of global fish exports (FAO, 2020). The least researched fisheries, and that have the least management in place, are being increasingly exploited.

Given the scale of China's population, length of its coastline and sheer number of people dependent on living marine resources for their livelihoods, China faces some unique challenges. However, the Chinese government has recognized the imminent challenges and has demonstrated strong political will for improving marine living resources management and for building a "marine eco-civilization" through recent regulatory and policy measures. There is much that can be gained from examining approaches and solutions from other nations, which have shared many of the same fundamental challenges in the management of living marine resources. Examining how different approaches and systems were designed in other countries, as well as understanding what went well and what could be improved upon, can offer valuable lessons and principles that China can certainly consider or apply to its own management. In addition, China can increasingly position itself as a thought leader in sustainable use and management of living marine resources, utilizing mechanisms such as the Belt and Road International Green Development Coalition to share its valuable technical expertise and experiences in living marine resource management with other nations.

The following case studies demonstrate ways in which other countries have found opportunities to enhance management of living marine resources in the areas of strengthened monitoring, holistic and integrated spatial planning, promoting long-term value over volume in marine fisheries, and scaling and adapting to proactively manage for a changing world to address complex issues like climate change. These stories can provide valuable experiences and ideas to draw from as China forges its path towards ecological civilization.

7.1 Strengthening Monitoring in Support of Governance

When living marine resource management is attempted without accurate or complete information, challenges inevitably emerge. In the context of living marine resources, monitoring can involve data collection to improve the performance or management of aquaculture operations, fisheries or ecosystem health. Monitoring can help indicate the success or failure of management, sometimes exposing unexpected patterns of information, which can help advance the state of the science of a particular issue. In addition, monitoring can also help detect non-compliance with rules and regulations, which can lead to stronger enforcement. Therefore, while increased efforts in monitoring can lead to better management, decision making and outcomes, lack of capacity and funding often poses a barrier to expanding such efforts. The following case studies demonstrate the value of monitoring and explore innovative ways to conduct monitoring to advance living marine resource management.

7.1.1 Information Transparency Facilitates Aquaculture Governance in Norway

Collecting, analyzing and sharing data on different marine production systems can help operators, fishermen, businesses, managers and other stakeholders make more informed decisions about living marine resource management. One example of information being collected and used to improve management comes from salmon farming in Norway.

Norway is the largest exporter of aquaculture products in Europe, and the sixth largest globally. However, given Norway's geography, 80% of the population lives <10 km from the sea, making it crucial to minimize negative impacts from aquaculture and to ensure that the ecosystems are well monitored. For salmon farming, the Norwegian Fisheries Directorate has a website that contains not only the location, capacity and operational status of every facility the government licenses, but also the environmental impact assessment results of each facility. All licenses for salmon farming in Norway are registered with this Geographic Information System (GIS), and each farm is required to submit weekly reports to the fisheries authority through the website. Government inspectors make routine checks on farms on a monthly basis. This near real-time sharing of data greatly facilitates mariculture governance and enables scientists to predict the environmental impact of facilities, such as the spreading of sea lice (Fig. 7.2).

The geographic information system of salmon farms on the Fisheries Directorate website has 13 thematic layers, covering information on localities of all aquaculture farms, the registered area of each farm, the place for fastening the net cage, biomass of cultured species, escapes, environmental condition, fish disease, coastal fishing data, various fjord data, maritime traffic, the Water Framework Directive, weather and wind, and administrative information (Liu et al., 2021).

Making data publicly available helps stakeholders understand government decisions, facilitates trust between the groups and creates a strong system of accountability. Scientists can review the data and propose better solutions, and the owners and operators of salmon farms are able to observe nearby environmental issues or disease outbreaks and get prepared accordingly. By gathering more data, the government is able to obtain a more accurate picture of the issues affecting, and resulting from, aquaculture and use that data to improve management overall.

Smart aquaculture technologies are becoming more and more popular in China in recent years, with some provincial fishery administrative departments and seafood companies setting up pilot-scale aquaculture data collecting systems. While data collection is an excellent starting point, and such efforts should be continued and expanded, data collection systems need to be designed with goals in place. Data should not be collected merely for its own sake, but should be gathered to answer specific questions related to ecosystem health and management effectiveness. Ensuring that data actually can generate new insights on the performance and impacts of aquaculture operations, which will then lead to better informed management decisions, is paramount. In addition, adaptive management mechanisms and

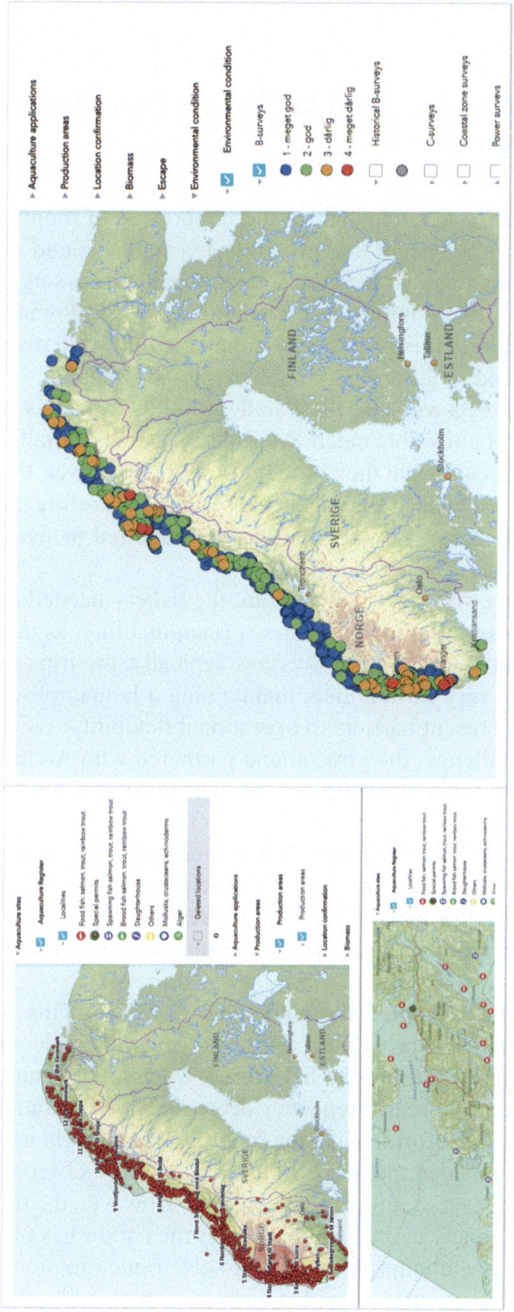

Fig. 7.2 Aquaculture farm production and environmental data published on the Norwegian fisheries directorate website (*Source* https://kart.fiskeridir.no/)

legal frameworks should also be in place, to allow the existing management measures being timely renewed and optimized.

7.1.2 Electronic Monitoring in the United States Pacific Whiting Fishery

Even when fishers and managers recognize the importance of monitoring and data collection, implementing a system can prove to be expensive and resource intensive. Utilizing technology to assist with the collection and processing of information offers great potential in overcoming such challenges. The following study shows how technology can address this problem by supplementing, and even supplanting, traditional observer-based monitoring systems in marine fisheries.

Comprised of around 35 vessels, the Pacific whiting fishery is managed with an annual quota for total allowable catch (McElderry, 2013). Initially, vessels were only accountable for the catch that they brought to shore. However, this incentivized greater offloading at sea of undesirable catch and bycatch. Therefore, the government revised the system and created new regulations that required fishermen to account for all catch, including fish discarded at sea.

To ensure compliance with this new system, the fishery needed a new comprehensive monitoring system that included at-sea accountability, as the nature of the fishery was not conducive to human observers. Typically, the trips for whiting are very short, and there is very little notice; maintaining a human observer would be very cumbersome, and present barriers to operational flexibility.

To overcome this challenge, the government partnered with Archipelago Marine Research to design an automated electronic monitoring system. With the electronic monitoring system, boats were equipped with up to four video cameras, fishing gear sensors and a GPS receiver. The data collected by these tools were stored on the boat, and recorded on a hard drive. This hard drive was retrieved by technicians every two weeks, and then sent to a processing facility for review. Meanwhile, fishers were still required to keep logbooks as a means for comparison with the automated data (Lowman et al., 2013).

Technicians regularly examined the automated data, searching for unreported discards and fishing activity in prohibited areas, and generally collecting vital statistics about the fishery that would prove helpful for fishers and managers. Over the seven years of the program, discard frequency decreased by 90% and data collection success increased to 98%. Unfortunately, the program was brought to an end in 2011 after the government mandated that the fleet had to use 100% observer-based monitoring as part of a broader set of reforms in the region. However, the requirement that a human observer accompany every boat that leaves the harbor has proven costly for the industry, which is now attempting to revert to electronic monitoring once more.

Unsurprisingly, human review of video footage can provide challenges itself, in the high cost involved with having a person sit and watch video and interpret what they are seeing. New approaches are being developed and piloted that involve machine learning, in which computers are programmed and trained to review and identify patterns in the video to identify when nets are being set, and when nets and fish are being brought in. This approach can greatly optimize the collection of data and film that is captured, as the camera would only be turned on when a programmed fishing event triggers recording. Therefore, this can significantly reduce the amount of footage a human observer has to sift through. Additionally, as machine learning and artificial intelligence become more and more sophisticated, there could be opportunity for computers and machines to do complete and full verification of catch, species identification, and other data collection and analysis functions to support fisheries management.

Fishing vessel positioning systems have been in operation in China in recent years, which have greatly facilitated law enforcement during the summer moratorium. Such tracking systems can also be utilized to ensure compliance with any spatial fishing restrictions. Electronic monitoring could help government authorities understand how many fish are being caught, in real- or near real-time. Collecting this information can serve to ensure compliance with any regulations on size or quantity of fish caught, as well as to provide data that support fishery science. At present, technologies exist globally that are appropriate for fishing vessels utilizing a wide range of gears, including trawls, purse seines, gillnets, pots or traps, and longlines (van Helmond et al., 2019). It is expected that continued advancements in machine learning and artificial intelligence positioning will continue to expand the range of functions and contexts electronic monitoring can be applied to, such as higher resolution monitoring and data collection of catch in large-volume, multi-species fisheries. Advanced imaging and processing technologies based on artificial intelligence should also be developed based on management needs in China, which may include detecting which species are being caught, where and at what quantities, all of which could help contribute to science and management needs, including implementation and expansion of output-based management systems, such as total allowable catch (TAC) programs.

7.2 Holistic Understanding and Integrated Planning

Holistic management of living marine resources will require the need to evaluate and balance tradeoffs between competing activities within the marine space. Research and science underpinning spatial planning, ecosystem services, proper aquaculture density and other activities can greatly improve the information that is used to make decisions about spatial planning. Combined with a robust planning process that involves multiple agencies and stakeholders, this holistic approach to management can help ensure that the benefits from sea uses can be maximized, while minimizing environmental impact and conflict among stakeholders. Below are several examples of how holistic and integrated management have been utilized for living marine resources in other countries.

7.2.1 Ecological Prioritization in Norway

Norway has recently finished designing and implementing a comprehensive system for the coordinated management of all sea uses. The ocean area of Norway is six times larger than its land area. In its marine space, a multitude of economic activity takes place, including fisheries, maritime transport and petroleum and energy. Naturally, conflicts can emerge among different resource users. For example, Norway's oil and gas industry experienced conflict since production began in the North Sea in 1971. There was opposition from fishers, who protested and claimed that the oil and gas companies were stripping them of their access to fishing grounds. The general public also worried about the presence of the oil and gas industries and how it would impact the health of their marine resources; this worry only intensified in the late 1970s as they faced their first major oil spill.

By the end of the 1990s, the Norwegian government had taken action on preserving its marine spaces, but was facing difficulties with managing different sectors. Beginning in 2001, Norway began to implement a single integrated management plan for one portion of its waters (Otterson et al., 2011). This effort was motivated in part by a newly passed Water Resources Act that ensured river systems and groundwater were managed in accordance with interests in the community. To create the plan, the government first worked to determine the biological, social and economic basis of the ecosystem and all activities that occurred in the space. The government then conducted impact assessments for all use sectors to see how the various activities could interact with or impact each other and the larger ecosystem (Pettersen, 2015).

This first plan, covering the Barents Sea-Lofoten (BSL) area, took four years to develop, and utilized a largely ad-hoc approach to its methodology. In the first stages of development, a steering committee was created from the Ministry of Fisheries and Coastal Affairs, Ministry of Petroleum and Energy, and Ministry of Foreign Affairs [in 2005, the Ministry of Labor and Social Inclusion and Ministry of Trade and Industry were added]. The steering committee was tasked with coordinating the management planning by setting overall aims and objectives, and delimiting which areas would be overseen under the BSL plan. Through aid from the Norwegian Polar Institute and Institute of Marine Research, ecological knowledge was translated into several maps, which were then analyzed with Geographic Information System (GIS) tools to highlight four "highly vulnerable areas" (Knol, 2010). The BSL plan covered 1,400,000 km^2, and these areas of high ecological vulnerability were prioritized with the intention of creating a management plan that would allow for coexistence of different sectors and the maintenance of the structure, function and productivity of the local ecosystems.

After these objectives and areas were set, the steering committee moved to Phase 2, or the analytical phase of the integrated management plan. During this phase, research institutions from related governmental directorates performed environmental impact assessments in delineated areas to determine the impact of various sectors, such as fisheries and marine transport. The assessments were also used to determine the impact of external factors, such as climate change and pollution, on the marine space.

In order to maintain a level of compatibility among each of the four assessments performed, common variables were set to facilitate comparisons. Upon completion, the assessment findings were consolidated into documents that then underwent a major peer review, with public hearings available for the general public. In addition, Internet discussions allowed stakeholders from the business sector, NGOs, local and regional authorities, and academia to post comments and provide feedback (Knol, 2010). Following public consultation, the steering committee created a *fagruppe,* or expert group of government directorates and research institutions, which was tasked with aggregating all the assessments from Phases 1 and 2 to set the vision, strategic goals and management objectives for the integrated management plan. The *fagruppe* ultimately proposed objectives centered on the themes of biological diversity and socio-economic goals of value creation, and mitigation of pollution and environmental risks. The process was then replicated in both the Norwegian Sea and the North Sea for a total of three marine plans. Each plan aims to produce greater value in these ecosystems, by pursuing the sustainable use of its resources while simultaneously upholding the integrity of natural ecosystems (Schive, 2018) (Fig. 7.3).

Setting a precedent in Norway, this ecosystem-based approach to an integrated management plan faced several obstacles. As to the ecological impact assessments (EIA), the actual intent or value implied of a resource was not standardized, so it differed among the assessments, even when looking at the same resource category. Additionally, since the EIAs were ranked on a five-point scale, from insignificant to catastrophic level of impact, what one assessment classified as minor might not have been classified in the same manner in a different assessment framework, and therefore presented elements of subjectivity (Ottersen et al., 2011). This problem was partially resolved when implementing the Norwegian Sea plan, which offered more standardization in scoring. During Phase 2, the researchers also found a significant lack of knowledge concerning seabirds, conditions of the sea beds, and trends in

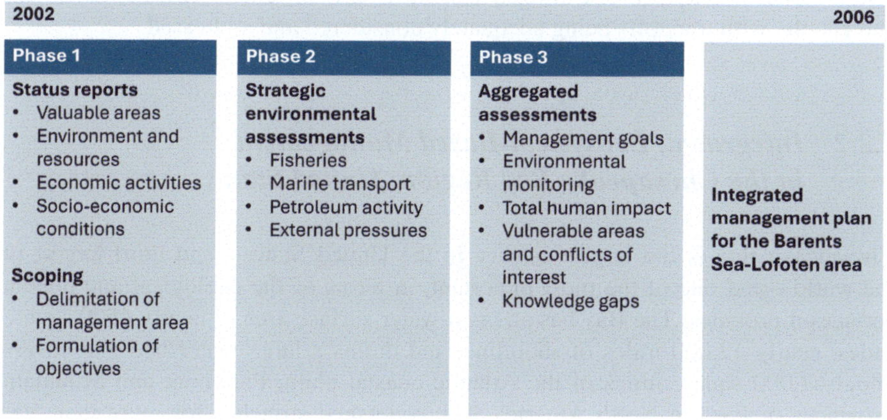

Fig. 7.3 Timeline of three stages of Barents Sea-Lofoten Plan (*Source* Olsen et al., 2007)

the levels of environmental hazardous substances and their impacts (Knol, 2010). This realization led to the increased funding of two federal programs, MAREANO (a program dedicated to mapping depth, sea conditions, habitat types and pollution in Norwegian marine areas) and SEAPOP (a program created to study seabird populations), aimed at decreasing the existing knowledge gap. For the sake of the plan, this knowledge gap also led the steering committee and researchers to create ecological quality objectives that would identify key components of the ecosystem and determine biological indicators in the sectors of pollution, fish resources, marine mammals and seabirds, and bottom fauna and habitats.

To balance these sustainable resource uses with ecological conservation, the planning and management process involved a number of government agencies, a high level of input from stakeholders, and intensive scientific study. One particularly valuable approach for managing such a complex process has been to establish ecological priorities from the start. The main criteria for prioritization are value—based on productivity or biodiversity—and vulnerability, based on concentration of all organisms, key life stages taking place, abundance of sessile organisms, migratory routes, etc. (Winther, 2018). These key conditions comprise the ecological foundation which the other uses must work around, and must not adversely impact. While each marine use is managed individually on a day-to-day basis, management is integrated into a higher level of governance which is bound by these ecological priorities, allowing the system to operate within a broad and connected vision. Through this ecosystem-based approach, Norway was able to not only develop and implement an integrated management system, but also construct a zoning approach based on vulnerability (Fig. 7.4) and successfully identify knowledge gaps.

The model and framework developed in Norway can provide several lessons for China, which has already begun to develop conservation areas and restoration plans. There could be potential opportunities for various agencies working on pollution, resource management and environmental monitoring to collaborate to manage larger systems. Through such collaboration, goals could be jointly articulated and a similar prioritization activity could be utilized to achieve optimal benefit for the environment and people, with tradeoffs being adequately considered and addressed.

7.2.2 Integrated, Ecosystem-Based Management in the Chesapeake Bay Region, United States

Chesapeake Bay is the largest estuary in the United States—and third largest in the world—and one of the most important, in terms of the ecological and human services it provides. The Bay has an open-water surface area of nearly 4500 square miles, nearly 12,000 miles of shoreline, and drains a large watershed that covers about 64,000 square miles of the Atlantic coastal plain, Piedmont and Mountain provinces of Eastern North America. The watershed stretches for more than 500 miles and includes parts of six states—Virginia, Maryland, Pennsylvania, Delaware,

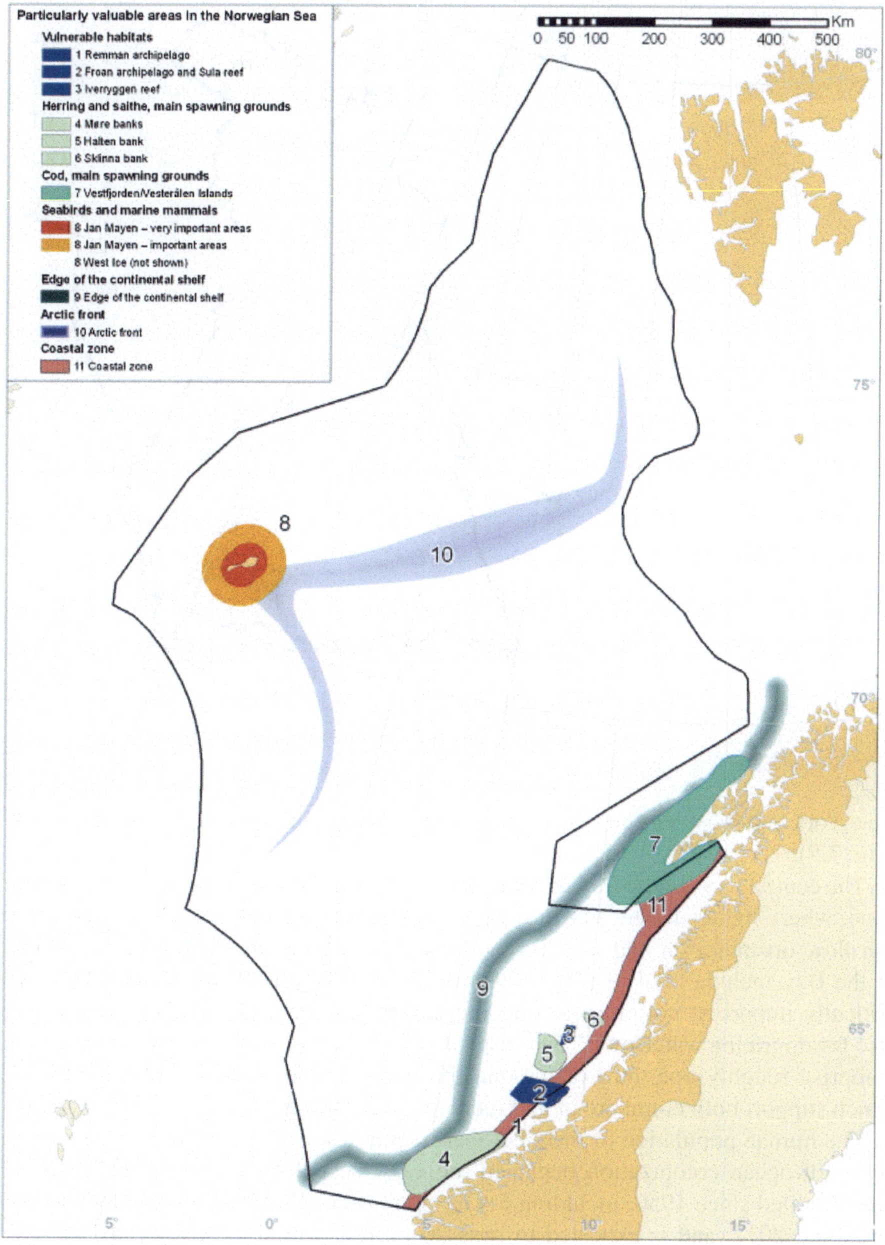

Fig. 7.4 Valuable areas within the management plan region (see legend for description) (*Source* Ottersen et al., 2011)

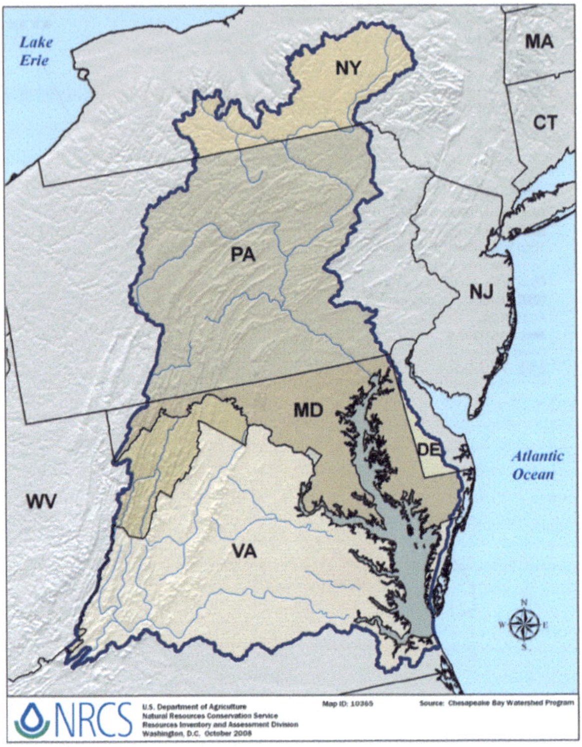

Fig. 7.5 Chesapeake Bay Watershed (*Source* USDA NRCS, 2008)

New York and West Virginia—as well as all of the nation's capital, Washington, DC (Fig. 7.5).

The central parts of the Bay are brackish—mesohaline—with an extensive mixing zone, where lighter freshwater moving downstream overruns denser saltwater that can flow upstream on and near the bottom. Key habitats in the brackish portions of the Bay include tidal marshes, seagrass beds and oyster beds, all of which are critically important not only for resident and migratory fishes and shellfishes, but also for migrating waterfowl. The most valuable fisheries focus on blue crabs, which comprise roughly one-third of national production, and oysters and striped bass, which support both commercial and recreational fisheries.

The human population of the Chesapeake Bay watershed has grown extensively since European colonization began in the early seventeenth century. It has more than doubled since 1950, including a 43% increase to 18.2 million people between 1980 and 2017, and is expected to pass 20 million within 15 years (CBP, 2012). As a result, the ecosystem has been highly altered by heavy land use on all sides (i.e., urban, residential, agricultural), as well as extensive in-water uses (e.g., fishing, boating, shipping, etc.). Over time, this has led to serious deterioration in water quality and fishing prospects. For example, excessive nutrient loading from both point and nonpoint sources—mostly sewage plants and farms, but also delivered through the air from wet and dry deposition of NOx emissions—led to blooms of microalgae, shifts in phytoplankton community structure and depressed bottom water oxygen levels (CBP, 2012). These anoxic events in the past have led to extensive fish kills—all over the region's estuaries—and periodic blooms of toxic dinoflagellates that can also kill fish and other aquatic animals. Over time, these impacts have worked to severely deplete oyster beds, seagrass beds and coastal wetlands, exacerbating direct effects of intensive fishing on Chesapeake Bay resources.

Even when stakeholders and governments noticed these environmental challenges, the large number of authorities involved on the land and in the water (e.g., private landowners, municipalities, states and the federal government) made it exceptionally difficult to implement and enforce overarching policies. For example, different environmental laws, such as the *Clean Water Act*, the *Coastal Zone Management Act* and the *Magnusson-Stevens Fisheries Management Act*, all serve to regulate activities occurring in the Chesapeake Bay. Each act focuses on different, but often intersecting issues, and is administered by different agencies. For instance, the U.S. Environmental Protection Agency (EPA) sets standards for pollution discharge permitting and water quality management, but delegates implementation to individual states. Meanwhile, the portion of the *Clean Water Act* focused on wetlands protection is still administered by the EPA, but implemented by a separate agency—the U.S. Army Corps of Engineers. The *Coastal Zone Management Act*, focused on land use planning in coastal zones, delegates management to states and local jurisdictions. The *Magnuson-Stevens Fisheries Management Act*, focused on the management of fishery resources, is implemented through an entirely different part of the federal government (the Department of Commerce), which defers management design to regional fisheries councils made up of state representatives, while still holding approval power over final decisions.

As one can observe, the governance mosaic and the array of agencies which provide resources for management are much more complicated and can lead to silos, which make integrated resource management extremely difficult. To work towards cohesive management across jurisdictions, interstate compacts and commissions have formed. For example, the fishery resources of the nearshore waters of the Atlantic coast are additionally managed by an interstate compact commission authorized under federal law, called the Atlantic States Marine Fisheries Commission (ASMFC). The ASMFC not only specifically manages most of the valuable finfishes of the mid-Atlantic (including all of the anadromous species, as well as most others), but is a vital sounding board and action agency for key cross-cutting challenges, including mitigation of climate change impacts on fisheries management. Another example is the Potomac River Fisheries Commission, which is a bi-state agency that regulates all fishing on the tidal main stem of the Potomac between the District of Columbia and the Chesapeake Bay, coordinating management between Virginia, Maryland and the District of Columbia.

The formation of the Chesapeake Bay Program is a key example of integrated management. In response to a grim Congressional research study that identified excessive nutrient loading as the cause of a steep decline in the living marine resources, the Chesapeake Bay Program was formed in 1983. Specifically, the program is a multi-jurisdictional, multi-stakeholder partnership established to coordinate policies, funding and technical capacity, and set ambitious, quantifiable goals with deadlines. It was founded as part of an amendment to the *Clean Water Act* (Sect. 320), which created a National Estuary Program (NEP). The NEP provides a coordinated function across federal programs and across stateliness within watersheds for integrated system-scale management—without impeaching the integrity of the individual component programs, or the authorities of the respective state and federal governments and agencies. All coordinated NEPs are adopted through *Comprehensive Conservation and Management Plans* (CCMPS).

More than 40 years ago, the Chesapeake Bay was the first estuary in the United States to be singled out for integrated protection planning, as obvious signs of nutrient pollution spurred action by the more affected states. In the 1970s, Congress dedicated US$27 million to a five-year study of needs and challenges, which was published in the early 1980s and lead to the creation in 1980 of the Chesapeake Bay Commission (initially a bi-state compact between Virginia and Maryland, later expanded to include Pennsylvania in 1985), and the 1983 signing of the *Chesapeake Bay Agreement* by the EPA; the states of Virginia, Maryland and Pennsylvania; the mayor of the District of Columbia; and the Chesapeake Bay Commission (CBP, 1983). The designated representatives of those institutions comprise today's Chesapeake Bay Commission, which serves as the legislative body for the overall program.

The initial focus of the Chesapeake Bay Program was to coordinate actions to redress nutrient pollution and its ecological, social and economic effects. In 1987, the *Chesapeake Bay Agreement* was revised to reflect the first numerical nutrient control targets; namely, a 40% reduction in nutrient loading by 2000 (CBP, 1987). This multi-state agreement was unique at the time, but it has since become common practice to establish specific numerical goals for priority outcomes with specific deadlines for action. In a 1992 revision to the *Chesapeake Bay Agreement*, the CBP began looking to reduce nutrient pollution sources more aggressively, and also to expand its area of interest to include toxicants. The second important expansion occurred in 2000, when the upstream states (New York and Delaware) committed to help reach goals; the final state, West Virginia, joined in 2002 (CBP, 2000).

Despite significant investments in coordinated action, impacts were less than what had been hoped for. Notable progress was made in land conservation, establishment of forest buffers from streams, and reopening fish passage blockages (for anadromous species), but the returns downstream in the Bay (in nutrient impacts and also on oysters) lagged. In 2009, President Obama reemphasized the importance of integrated conservation, signing an executive order directing federal agencies to give it their full attention.

By 2010, the EPA had worked with all seven jurisdictions to develop and adopt a formal and legally binding nutrient loading target, called a *Total Maximum Daily Load* (TMDL), with obligations for load reduction (i.e., from approximately 284 to 200 million pounds of nitrogen, and from 16.3 to 15 million pounds of phosphorus per year from all sources) that must be achieved throughout the Bay by 2025. Each jurisdiction has now adopted a *Watershed Implementation Plan* (WIP) to meet those objectives. Taken together, the TMDL and the WIPs are sometimes referred to as the *Chesapeake Bay Blueprint*. The most recent reauthorization of the *Chesapeake Bay Agreement* was in 2014; for the first time, all necessary jurisdictions agreed on management to achieve 10 specific goals and 31 specific outcomes, from water quality to habitat to living resources (CBP, 2019). Private interests delayed full implementation of the *Chesapeake Bay Blueprint* in the lower courts from 2011 to 2016, but the U.S. Supreme Court refused to reverse the courts' decision, and the agreement remains in effect today.

Today, many impacts from unchecked development and utilization remain. Nevertheless, measurable improvements in water quality and habitat, and oyster, blue crab and other wildlife populations have been achieved through reducing pollution and other adverse impacts, protecting healthy habitats where possible and widespread restoration efforts. While the most important contribution of the CBP is in coordinating planning and action among numerous agencies and institutions around key environmental targets, the program is also responsible for maintaining monitoring and assessment systems related to Chesapeake Bay. These efforts include data collection, analysis and reporting against priority goals, which help minimize conflict and support decision making among a broad set of stakeholders and jurisdictions with competing needs or interests.

Still, even this large-scale effort falls short in meeting the management needs of some very highly migratory marine and anadromous species that call the Bay home. Full success requires coordinated actions through integrated fishery management planning and habitat protection planning at an even broader scale. Thankfully, other programs are also in place that can help achieve those broader goals. For instance, integrated management occurs at both a coastwide scale for nearshore fisheries through the ASMFC, and for essential fish habitats through the work of the Mid-Atlantic Fisheries Management Council (and NOAA). Both of those federal and interstate processes are tightly focused on ecosystem-based management of nearly all key species not adequately tracked at the Bay scale.

An emerging issue that the CBP could be well poised to tackle, with the help of other multi-jurisdictional institutions, is to address climate impacts on this ecological system and the human services it provides. Already, species are shifting northward, some relatively rapidly. For instance, one key anadromous species, alewife (a river herring), may soon no longer be able to access its major historic spawning grounds in Albemarle Sound, just to the south, with uncertain effect on coastwide spawning potential; more warm temperate species are moving in. Those shifts would make current allocations of access for fisheries obsolete, and even strain the way governance is organized. In addition, shifting precipitation patterns will affect nutrient delivery patterns and trends, and potentially induce ecological cascades within the Bay itself; salinity and alkalinity shifts could exacerbate those effects. Many scientists are concerned about already long-term shifts in phytoplankton and zooplankton communities in the Bay, induced by nutrient shifts, that have replaced larger and more available phytoplankton with nano- and picoplankton.

The CBP demonstrates one potential strategy China might consider for managing increasingly large and complex challenges that come with mixed use of natural resources, specifically in areas where there are dense urban populations or multitude economic activities heavily impacting natural systems. Many of the key pieces for integrated management, such as extensive environmental monitoring programs, pollution control efforts, habitat restoration and habitat reconstruction efforts, are underway. With emerging challenges from issues like climate change, there could be opportunities to build from this fundamental infrastructure. For example, there could be opportunities for various environmental and regulatory agencies to work more closely to develop comprehensive management plans that more holistically consider benefits, impacts and tradeoffs of different resource uses, and impacts to the natural environment. This model of integration and collaboration can perhaps even extend to more close cooperation between different jurisdictions, such as between provinces, to enhance information sharing, joint goal setting and larger-scale management. Deciding the best mechanism will depend on what is appropriate for China's unique needs, but the Chesapeake Bay Program demonstrates that separate—or altogether new—entities that bring together representatives of various states and agencies is one model of creating the basis for larger scale, more holistic and integrated management.

7.2.3 Spatial Management on the Great Barrier Reef

The Great Barrier Reef is popularly known for containing the largest coral reef ecosystem in the world, but it is home to several other marine habitats, such as seagrass, mangroves, sand, algal and sponge gardens. During the 1960 and 1970s, a particularly destructive outbreak of corallivorous crown-of-thorn starfish infested the Great Barrier Reef and severely impacted the health of the coral reef ecosystem, so much so that in 1981, surveyors estimated that approximately 90% of the hard coral cover of the reef had been killed (Endean, 1982). This starfish infestation, coupled with concerns over increasing damage from petroleum mining, led Australian legislators to take action to protect a natural wonder and cultural icon from ongoing environmental threats. In 1975, a federal law created the Great Barrier Reef Marine Park, a multi-use park designed for fishing, recreation and shipping, among other sectors, comprised of smaller sections, each with its own zoning (Gershman et al., 2012; Fig. 7.6). However, throughout the 1990s, scientific evidence began to show that this piecemeal approach was not effective. The initial spatial management plan involved implementing partial zoning, a method that splits an individual area into one zone, with the rest of the area marked as a different zone. With more than 2400 individual reefs within the Great Barrier Reef, this inconsistent zoning led to challenges in public understanding, as one reef could be open for public access and all activities on one side, and be restricted to non-extractive activities on the other. Partial zoning also made enforcing non-compliant behavior difficult, as the close proximity of zones made it difficult for enforcement agencies to distinguish where activities were taking place. Under the 1975 act, 95% of trawlable area was still available for commercial trawling, with 75% of the seabed trawled each year; it was not until the *Trawl Management Plan* of 2000 that action was taken against the negative impacts of

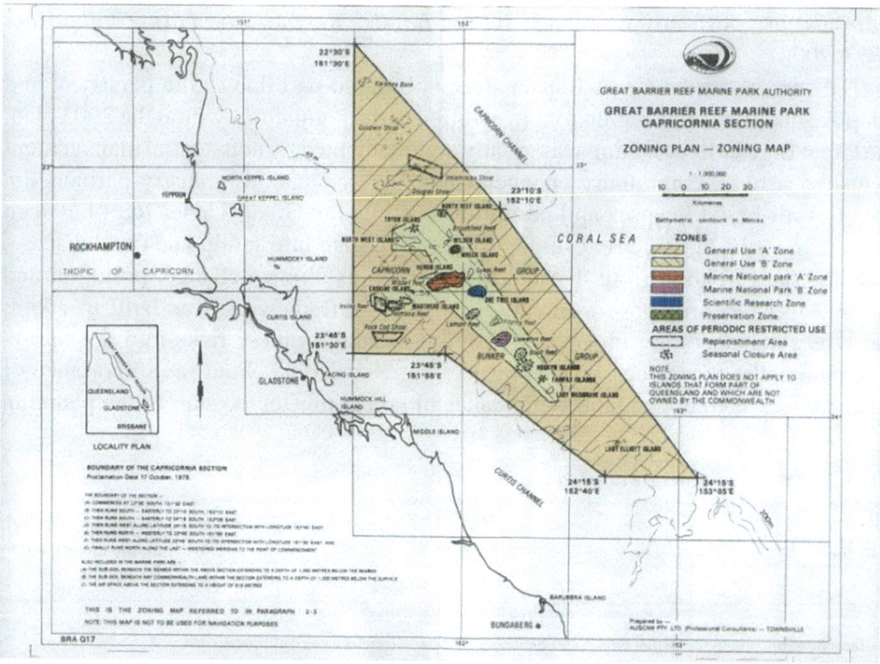

Fig. 7.6 Example of zoning sections—Great Barrier Reef Marine Park Capricornia Section zoning plan, August 1980 (*Source* Gershman et al., 2012)

the trawling fishery on the biomass of epibenthic organisms (Craik, 1992). It became apparent that the 1975 plan, while incredibly ambitious, failed to properly account for the impacts of the fishing industry and the issues that partial zoning would create for compliance and enforcement.

Instead of continuing with an ineffectual approach, the government reexamined the evidence and began a new strategy. A comprehensive and science-based planning process was launched to define representative areas, covering the range of habitats, species and communities across the entire park, contrary to the previous approach of first breaking the park into sections. It was a long, participatory and consultative process, heavily involving stakeholders and community consultation and participation from the initial classification up until parliamentary approval (Fig. 7.7). Authorities considered social and economic impacts and sought alternatives that met environmental goals while minimizing adverse impacts. While the process was lengthy, it was crucial to investigate all factors and consider all options, so that the design proved more effective and more durable in the long term. The re-zoning plan, finalized in 2004, protected 70 different bioregions and increased no-take zones from 4.5 to 33%, providing an adaptive buffer for change as a result of unexpected disaster (Evans et al., 2014; Fig. 7.8). This re-zoning plan spawned nine different zoning areas with varying degrees of use regulations, from general use zones that allow "all reasonable uses" except for mining and drilling, to preservation zones, where all entry and use is strictly prohibited (*Great Barrier Reef Marine Park Zoning Plan 2003*, 2003).

The success of the Great Barrier Reef is in large part due to the persistent and adaptive qualities of its regulation. In addition to the regulations within the 2003 plan, the Great Barrier Reef zoning was greatly complemented by non-spatial management of fisheries (e.g., community engagement, public education, industry partnership, fishery skill-based licenses and permitting). While the Great Barrier Reef has seen success in terms of decreased susceptibility to starfish infestation and coral disease, and increases in coral trout density, threats such as climate change, pollution and natural disasters still pose a threat. However, the framework was built to adapt; therefore, managers are in a better position to address these risks into the future. Internationally, the process stands as a valuable anecdote about the importance of adaptive management and the significance of taking time at the start of the planning

Phases of the Representative Areas Process

Phase	Description	Stakeholder Involvement and Communication
Classification	The Representative Area Program maps the diversity of the Great Barrier Reef World Heritage Area into bioregions using biophysical data and expert opinion.	✓
Review	The GBRMPA determines the extent to which the existing zoning protects the biodiversity shown by the bioregions.	✓
Formal Community Participation Phase 1	Advise the public of the proposal to review the existing zoning to better protect biodiversity and invite comments on the sorts of information to be considered when developing a new representative area network.	✓✓
Identification	Extensive community consultation to identify networks for potential inclusion (e.g., candidate areas) which will achieve the objectives of RAP and could be incorporated into a new representative area network.	✓✓
Selection	Extensive community consultation to assist in selecting among the options of networks of candidate areas. Assessment of social, economic, cultural, and management implications to assist in choosing the most appropriate candidate sites and levels of protection.	✓✓
Draft Zoning Plan	Draft Zoning Plan is prepared, taking all community consultation into account.	✓
Formal Community Participation Phase 2	The Draft Zoning Plan is made available for public comment.	✓✓
Development of Revised Zoning Plan	Development of a revised zoning plan taking consideration of all public comments.	✓
Ministerial and Parliamentary Approval	Ministerial approval and Commonwealth Parliamentary approval (15 sitting days) before it can become law. The new zoning plan comes into effect on a date chosen by the Minister.	✓

Fig. 7.7 Phases of representative areas process (2004 Re-zoning) (*Source* GBRMPA.GOV.AU)

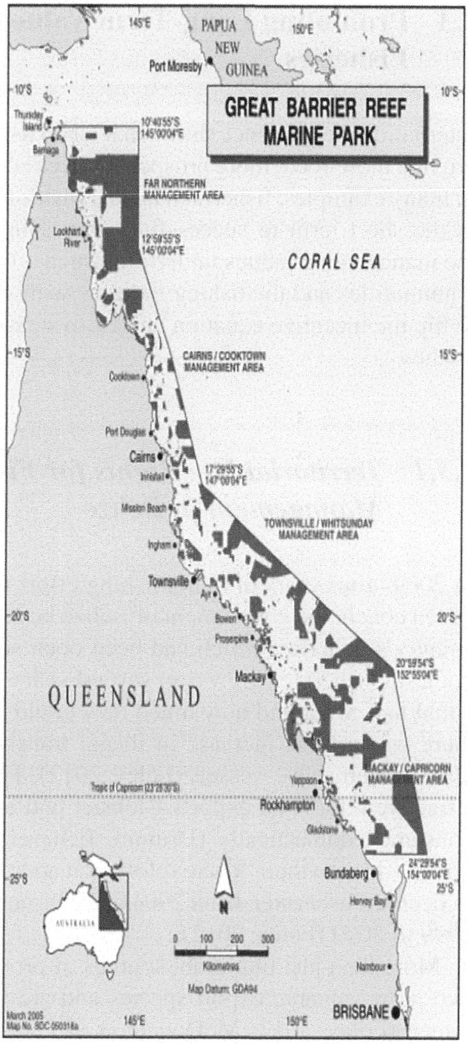

Fig. 7.8 Great Barrier Reef Marine Park—Location and size of New Marine no-take areas (*indicated in dark gray*) (*Source* Fernandes et al., 2005)

process to integrate as much information as possible, and involve a wide range of stakeholders and perspectives.

For China, the two biggest takeaways are the spatial elements of management that account for ecosystem dynamics, and the adaptive nature of management. China has already made progress in designing conservation areas to support biodiversity, and the Great Barrier Reef example shows the importance of measuring the success of conservation and spatial management in order to collect information and data, consider emerging threats and adapt existing spatial management plans as needed.

7.3 Promoting Long-Term Value Over Volume in Wild Fisheries

International experience shows that when fisheries are sustainably managed, they can provide more food, more prosperous livelihoods and a healthier marine environment. In many examples, fishermen have transitioned from maximizing their catch volume in the short term to successfully increasing the fishery's long-term value. While the management tactics underlying such a transition can vary, providing fishermen, communities and the fishing industry with strong access rights has proven effective to flip the incentive equation and cultivate a conservation ethic, as in the below case studies.

7.3.1 Territorial Use Rights for Fishing (TURF) Management in Belize

In 2009, after years of rising fishing effort and declining catch of spiny lobster and queen conch, the government of Belize began an effort to improve governance of the country's fisheries, which had been open access up until that time. The number of domestic fishermen grew over several years, without any distinction as to where they could fish, what and how much they could catch, among other factors. In addition, there was a large increase in illegal transboundary fishing by foreign—in particular, Guatemalan—vessels (Foley, 2012). The loosely managed access to water, and attractive returns on exports—lobster and conch are primarily exported rather than consumed domestically (Drumm, Echeverria & Almendarez, 2011; Fujita et al., 2017)—lead to their "overexploitation and decline", as evidenced by a 30% increase in licensed fishermen from 2004 to 2008, and a 24% decline in lobster landings from 1999 to 2009 (Foley, 2012).

More than just important sources of protein, lobster and conch are the country's two most valuable export species and are vital to the income of the nation's fishermen (Foley, 2012; McDonald et al., 2014). With such high stakes, the government decided to design and implement a Territorial Use Rights for Fishing (TURF) system, which is locally referred to as managed access. A TURF is an area-based fishing right approach that defines fishing area and allocates secure fishing access to groups or individuals. TURF pilot projects were launched by the Belize Fisheries Department in July 2011. They were established in two geographically adjacent, yet ecologically diverse, communities: Glover's Reef and Port Honduras, situated on the Southern Barrier Reef. The design process involved the community over 80 meetings, including information sessions on rights-based management, discussions of the managed access system and simulated design activities (Castañeda et al., 2011). Additionally, one-on-one and group meetings, as well as workshops with the fishermen, allowed stakeholders to build a shared agreement that managed access could help tackle the challenges of increasing competition for—and decreasing amounts of—catch, amidst rampant illegal fishing and limited organization of fishery resource management (Castañeda et al., 2011).

At the pilot sites, fishing was regulated or banned, depending on the specific area. In regulated zones, access was restricted to local fishermen who had fished in the area historically. These fishermen in the pilot areas were granted the right to fish in exchange for respecting no-take zones and other fishing regulations, including the requirement to report catch data. Managed access committees were incorporated, with their membership comprised of fishers representing fishing cooperatives or villages, which were meant to inform the Belize Fisheries Department's decision-making process on licensing, as well as update both the regulators and the communities on the progress of the managed access program (Fujita et al., 2017).

The results of the pilot projects included less competition on the water day-to-day and more investment in the future health of their resources (Casteñeda et al., 2011); arrests also dropped by 60% relative to the same time period in the prior year. The data at the sites was collected through independent monitoring, self-reporting by the fishermen (such as catch logs) and a surveillance system (Casteñeda et al., 2011). In 2013, improvements were observed, with 80% of fishers reporting their catch data—more than 70% of whom reported higher catches—as well as increased compliance with fishery regulations, such as less poaching activity in no-take areas and less fishing during the closed seasons for lobster and conch (Fujita et al., 2017).

The pilot projects served an important role in nudging the fishing communities to adapt to the shift from an open to managed access system; as such, the pilot projects were considered a success (Fujita et al., 2017), and the Belizean government scaled the program nationally in June 2016 to cover all domestic waters (Fig. 7.9).

As elaborated in Fujita et al. (2017), the TURF model was scaled to all of Belize's territorial waters, whereby fishers could fish in one to two managed access areas of their choice. Such perception, in addition to community consultation and a social media marketing campaign, allowed the roll-out of the nationwide managed access system (Wadea et al., 2019). Managed access committees, equivalent to those in the two pilot project areas, were established in the seven additional managed access areas. All conch and lobster fishing is now governed through a network of TURFs and co-managed among the national government and local committees comprised of elected fishers. Further, a revised data reporting system was also implemented (Wadea et al., 2019).

Despite the limited availability of comparable data, several developments since 2016 are worth highlighting. According to the Health Reefs Initiative, Belize's Southern Barrier Reef has improved since 2015 to achieve an overall "good" health rating, one of only three reefs in the Mesoamerican Reef to do so in 2018 (McField et al., 2018). Further, national fishery management plans for conch and lobster were adopted and, in November 2017, queen conch catch limits (i.e., total allowable catch for the season) were implemented by the Belize Fisheries Department (*The San Pedro Sun*, 2017). Additionally, April 2019 saw the no-take (i.e., replenishment) zones grow, expanding from 4.5 to 11.6% of total territorial waters area of Belize, contributing to the efforts to improve the situation for the species living in the managed access areas.

Fig. 7.9 Belize's national system of TURFs (*Source* Fujita et al., 2007)

In 2020, the Belizean government passed the *Fisheries Resources Bill*. The bill enables better enforcement of the fishery regulations and replaces the *Fisheries Act* of 1948 (Griffin, 2019; MacKenzie, 2019). The initial focus on lobster and conch is now being expanded to the high diversity finfish fishery through the application of data-limited tools. Fisheries management plans for multispecies finfish are also anticipated (Martinez et al., 2018). Fishermen are invested in and supportive of strengthening and expanding the network of marine protected areas, so as to improve the quality of their resources. To leverage the better product quality, the cooperatives have partnered with NGOs to create a local seafood certification program with Belizean restaurants and hotels (Fujita et al., 2019). This new market of local businesses opens the door to a premium price for a sustainable product and rewards fishers for upholding their sustainable practices in the long term.

The TURF, area-based approach is one that China may consider in specific contexts. TURFs are particularly well suited in situations that involve sessile species that do not travel great distances, such as crabs or other demersal species. Design features, such as closed areas, and integration of other marine uses, such as tourism, subsistence fishing and conservation areas, can be well integrated into TURFs. In addition, because TURFs assign specific fishing areas to groups or individuals, participants tend to develop a stronger motivation for conservation and compliance with rules. They know that if they are stewards of their designated fishing areas, they can benefit from them. This is in contrast to the open access scenario, where a multitude of different fishing groups tend to compete in a free-for-all type setting, and often carry a more short-term view. A well designed TURF that includes input and participation from the fishing community can allow policymakers and the fishing community to balance and achieve biological, economic and social objectives, for the benefit of both people and the environment.

7.3.2 ITQ Management of Red Snapper in the Gulf of Mexico

The Gulf of Mexico supplies more than 40% of the United States' domestic seafood. Among the catches of crab, shrimp, groupers and swordfish is the commercially and recreationally prized red snapper *Lutjanus campechanus*. The red snapper population underwent a rapid depletion starting in the 1950s, as a result of overfishing due to increasingly effective fishing gears, such as double-rig trawls, and mass production of fiberglass boats and motor technology, which allowed for higher commercial and recreational landings, respectively. Meanwhile, a booming shrimp industry began to contribute high amounts of red snapper bycatch. By 1990, spawning potential had declined to just 2%, compared to the target spawning potential of 26% (NOAA, 2018) (Fig. 7.10).

Recognizing a dire situation, management authorities attempted an aggressive series of reforms, including implementation of size limits, seasonal closures and TAC. However, the design and implementation of these regulations often failed to consider the incentives or operational needs of the fishers. For example, when the TAC limit for red snapper fisheries was lowered from five to four million pounds in 1991, an intense competition among commercial fishermen (often known as derby fishing) occurred. As a result, the fishing season closed halfway through the year, once the TAC was reached, and closed in just 53 days in 1992 (NOAA Fisheries, 2017). Such drastic shortening of fishing seasons severely limited fishing opportunities and reduced fishers' ability to most optimally meet market demand, as larger volume catches in shorter periods can lead to market saturation and lower prices. Specifically, shorter seasons made it difficult for dockside buyers to find demand for fish being landed in a narrower window of time, resulting in sharp declines in the price of red snapper at the beginning of each open season (Waters, 2001). Beyond economic

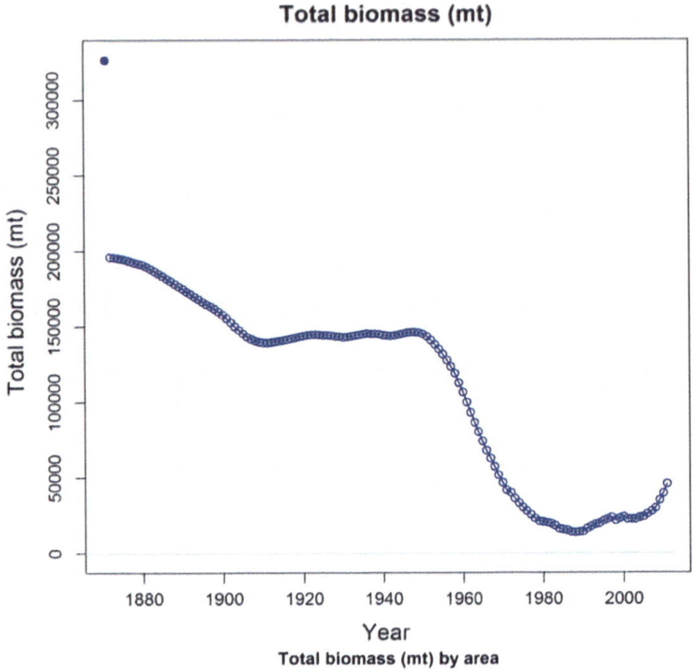

Fig. 7.10 Spawning stock predicted total biomass for Gulf of Mexico Red Snapper from the base model run (East & West data combined) (*Source* SEDAR Southeast data, assessment, and review SEDAR 31 Gulf of Mexico Red Snapper Stock Assessment Report, 2013)

considerations, the race to fish resulting from shortened seasons and lower catch limits created more pressure for fishers to operate under dangerous conditions.

A new approach was needed that better aligned fisher incentives with conservation, and was more suited to their operational needs. On January 1, 2007, the managers implemented an Individual Transferable Quota (ITQ) program to reduce overcapacity and eliminate the problems associated with derby fishing. Using historical landing data, ITQs were allocated to eligible participants, and a vessel monitoring system was introduced alongside real-time data management to hold fishers accountable to their quotas. The ITQ program consolidated the number of fishers on the water and assigned each of them an individual share of the fishery's TAC. The TAC is set annually by the Gulf of Mexico Fishery Management Council based on regular scientific stock assessments and with the advice of a committee of scientific and statistical experts. In 2007, at the start of the program, the TAC was drastically lower than in 2006 (Agar et al., 2014). In some cases, the harsh reduction and resulting small individual quotas caused hardship for fishers. However, by 2009, assessments indicated the stock was rebuilding and managers increased the TAC. In recent years, the TAC has been raised beyond pre-ITQ levels, due to the healthy recovery of the fish stock. As the program has continued, the number of fishing trips has decreased, the number of discarded fish has decreased, fish stocks have rebuilt, the catch limit has doubled and there has been a 100% increase in revenue for fishers. This revenue increase is due to the fact that fishers are now able to fish when market demand and prices are higher—for example, during the season of Lent, a Catholic holiday during which many Americans give up eating red meat and substitute for products like fish—and they are catching higher quality fish.

Although the implementation of an ITQ catch share program has helped the commercial sector move towards more sustainable fishing practices, a lack of cohesive management between the commercial and regional sectors presents further challenges. Whereas the commercial sector has not exceeded its quota since the establishment of the ITQ system in 2007, the recreational sector's season lengths have steadily decreased, down from a 40-day season in 2012 to a short, 10-day season in 2015 (Powers & Anson, 2016; Fig. 7.11). In 2015, the recreational sector split into a private angler sector and a for-hire sector, each with a defined share of annual catch limit (ACL), and managed as a derby-style fishery (NOAA, 2017). Despite efforts to reform the recreational sector, there has been difficulty among recreational fishers in considering their impacts on the long-term stability of the red snapper population (Powers & Anson, 2016). As these fishers struggle to deal with shorter seasons and decreasing daily take limits, while still regularly exceeding their quota, conflict has grown between the recreational and commercial sectors. In 2014, the commercial sector sued the National Marine Fisheries Service (NFMS) for not accounting the recreational sector's overfishing into their 2013 quotas. The NMFS subsequently shortened the 2014 recreational season from 40 days to nine days (Morrison, 2016).

Traceability, for purposes of management accountability and consumer awareness, is another important factor of sustainable fishing that is improving over time. Gulf Wild, an organization engineered by red snapper fishers in the Gulf of Mexico, ensures authenticity and responsible harvesting of their fish through a QR code (Quick Response code) on their packages that, when scanned, provides details for each individual fish (e.g., who caught it, where it was caught, etc.) (Gulf Wild, 2019). Consumers can then use this information to consciously choose to support sustainable harvesting of red snapper. This traceability and certification program allowed fishers to obtain better prices for their fish, and is one successful example of maximizing value of a species versus maximizing volume.

With many challenges still left to confront, it is important to understand the benefits of reform efforts. Since the implementation of the commercial red snapper catch share in 2007, the stock has more than tripled in size (NOAA Fisheries, 2013), and federal regulators have increased commercial catch limits by 300% since 2008 (NOAA Fisheries, 2017). Dockside values have grown from approximately US$10 million to US$28 million in the same period of time, and despite systemic overharvesting in the recreational sector, red snapper continue to rebuild (NOAA Fisheries, 2013; NOAA Fisheries, 2017).

This particular case study is relevant to China as it seeks to implement sustainable fisheries management in coastal waters through the TAC provincial pilot projects. As highlighted in the U.S. red snapper example, China may need to think creatively about how best to design and implement such programs to achieve its own biological, economic and social goals. As success is achieved in TAC setting and implementation,

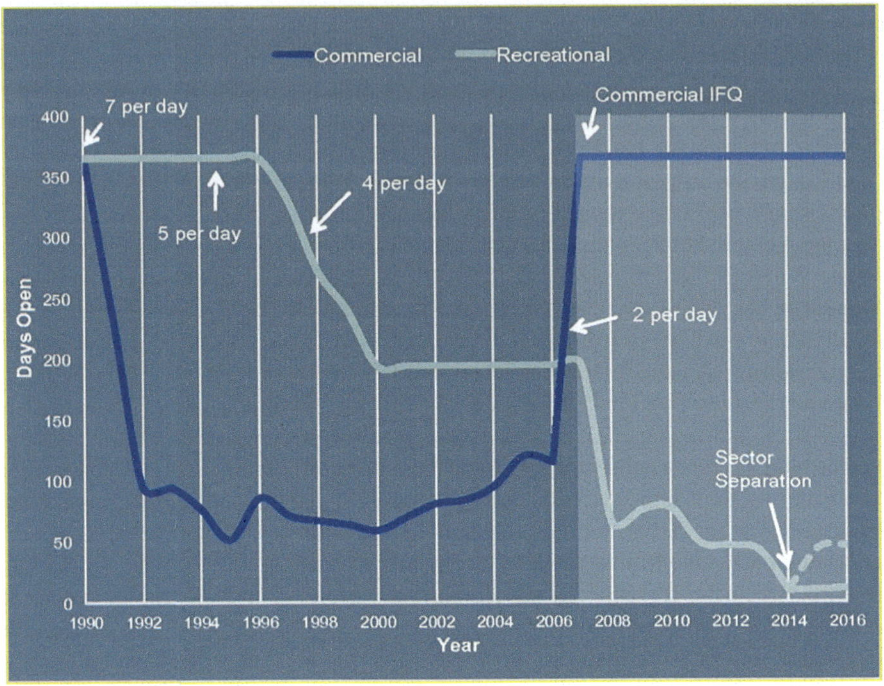

Fig. 7.11 Management history of the recreational and commercial Gulf of Mexico Red Snapper Fishery (*Source* Abbott et al., 2018)

it may be worthwhile to consider in which cases further allocating this quota to groups or individual fishers would offer additional benefits. The Gulf of Mexico example shows that individual allocations help align fisher incentives with longer term sustainability outcomes. When granted quota, fishermen have a stronger stake in the long-term health of the resource, which can motivate stronger compliance with regulations. Under such an arrangement, fishermen can feel more assured that their positive actions in the short term can benefit them in the long run. Conversely, without holding the individual quota, and by merely competing for a share of the TAC, this assurance may not be as apparent to the fishermen, as evidenced in the earlier years of U.S. red snapper management under only a TAC. In addition, the red snapper fishery example highlights the importance of considering market conditions in design of management; specifically, the need to prevent fishing derbies and races for fish by allowing fishermen to decide when and how much they harvest, based on optimal market conditions. Although individual quota systems take more effort to implement and monitor, they allow fishermen to operate with more of a business mindset, focusing on quality and fetching higher prices rather than trying to maximize quantity of catch, which can help them extract more value from the same amount of available resources. Such a balance can help fishermen continue to maintain or enhance their livelihoods while supporting the long-term sustainability of living marine resources.

7.4 Scaling and Adapting to Proactively Manage for a Changing World

Previous sections in this chapter focused on case studies that could support China's ambitions to create a domestic ecological civilization in the area of living marine resource management; however, many living marine resources, such as migratory species or those that are shared with regional neighbors, can often benefit from stronger coordination and integrated management across jurisdictions. These jurisdictional considerations are particularly prominent in the face of climate change, which has already begun to shift the geographic ranges of many marine species. In addition, with China's continued advancement, innovation and progress in addressing complex challenges in areas such as aquaculture, integrated spatial planning and marine fisheries management, there is tremendous opportunity for other countries that are faced with similar challenges to learn from Chinese models. This potential for collaboration can complement existing efforts and ambitions—such as China's Belt and Road Initiative—to support sustainable economic and environmental development regionally, thereby expanding the ethos and benefits of ecological civilization more broadly.

7.4.1 Managing for Climate Change

As presented in Chap. 4, the effects of climate change are already impacting, and will increasingly impact, all ocean ecosystems and every country's living marine resource management. Threats range from changes in stock distribution and productivity to questions of food security and fisher safety. However, the severity or combination of climate change impacts will differ across contexts. There is no universal solution to this problem. Some stocks will increase in some regions; others will disappear. Some mariculture operations may thrive from temperature changes; others may fail from extreme weather (De Silva & Soto, 2009).

Countries are already being confronted with these challenges. One contentious example is the so-called mackerel wars between the EU and Norway on one side, and Iceland and the Faroe Islands on the other (Fig. 7.12).

Mackerel is found across the North East Atlantic and was typically fished by Britain, Ireland, and Norway. Meanwhile, Iceland and the Faroe Islands typically fished for other species, such as blue whiting. Over time, as blue whiting became overfished, Iceland and the Faroe Islands started to target mackerel. At the same time, with rising sea temperatures due to climate change, mackerel began migrating further

Fig. 7.12 Map of Britain, Norway, Iceland and the Faroe Islands (*Source* British Sea Fishing, 2019)

north and became more plentiful in the EEZs of both Iceland and the Faroe Islands. As Iceland and the Faroe Islands discovered a greater number of mackerel in its water, they increased their quota significantly—from around 2,000 tons to 130,000 tons for Iceland, and from 25,000 tons to 150,000 tons for the Faroe Islands starting in 2010 (British Sea Fishing, 2019).

Both countries justified their large increase in national annual quota with climate change. They claimed the increased northward migration of mackerel due to climate change led to more plentiful mackerel in their waters, and that all additional fishing fell into their EEZs. However, the higher national annual quotas were unilaterally imposed without consultation and were against the EU-wide agreements in the Common Fisheries Policy.

In response, Britain, Norway, and Ireland felt a need to take direct action against Iceland and the Faroe Islands. In August 2010, Norway closed all ports to Iceland and Faroe Islands trawlers. One Faroese trawler *Jupiter* was forced to turn around and return to the Faroe Islands instead of unloading its mackerel catch. In 2011,

Iceland reignited protest by further increasing its quota from 130,000 to 146,818 (Henley, 2019). The EU condemned Iceland and the Faroe Islands for inflaming an already tense situation and warned that Iceland's bid for full EU membership would be negatively impacted by the mackerel war. Iceland and the Faroe Islands continued to justify their quota with the abundance of mackerel in their waters and their right to set their own national quotas, arguing that Europe needed to share quotas more equally. Iceland also began to ban Scottish trawlers targeting other species from Icelandic waters, which brought non-mackerel fishermen into the consequences of the mackerel war.

One of the sources of Britain, Norway, and Ireland's opposition was that the increase in Iceland and Faroe Islands quotas could lead mackerel to lose its status as fully certified as sustainable by the Marine Stewardship Council (MSC). With the loss of MSC accreditation, consumer demand for mackerel would decrease and Europe would lose one of its few sustainable fisheries. This came to fruition in March 2012 when the MSC officially announced that all North Sea and North Atlantic mackerel would no longer carry MSC accreditation because it did not meet sustainability requirements (Jensen et al., 2015). The MSC projected that mackerel catches would exceed 900,000 tons in the near future, which was at least 260,000 tons over recommended scientific guidelines. The suspended accreditation affected mackerel in Ireland, Norway, Denmark, Sweden, Britain and the Netherlands.

As a result, Iceland reduced its mackerel quota by 15% in 2013 and hinted that they might reduce the quota more if other countries did the same. However, the EU countries and Norway were unimpressed and continued to argue that Iceland and Faroe Islands quotas were unsustainable. Even with a 15% reduction, Iceland claimed almost a quarter of total mackerel catch. By 2014, the EU and Norway were prepared to place sanctions on Iceland and Faroe Islands fish products.

However, a partial resolution was reached in March 2014 through a trilateral agreement that split the mackerel quota between the EU nations and Norway (72%), Iceland and Russia (16%), and the Faroe Islands (12%). This resolution would not have been possible were it not for a revised assessment by the International Council for the Exploration of the Sea (ICES) that found an increase in the spawning biomass of mackerel. This meant there was more available mackerel for catch, so several countries actually saw their mackerel quota increase under the new trilateral agreement. After two years of MSC assessment and coordinated support from the Mackerel Industry Northern Sustainability Alliance (MINSA), in May 2016 the MSC restored the status of North East Atlantic mackerel as a sustainable species (Marine Stewardship Council, 2016). As a collaboration of over 700 fishing boats from eleven countries, MINSA represented international commitment and cooperation towards easing the mackerel wars.

There are still several underlying issues in the aftermath of the mackerel wars. The partial resolution only came about because mackerel stocks were found to be higher than originally predicted. Therefore, future reductions in mackerel stock could reignite another mackerel war as countries fight to maintain their current quotas. Furthermore, the North East Atlantic Fisheries Commission (NEAFC), which is made up of the EU, Norway, Denmark, the Faroe Islands, Greenland, Iceland, and

Russia, is supposed to be responsible for managing and enforcing the mackerel stock. However, Iceland, Russia, and Greenland do not collaboratively decide quotas and can effectively set their own limits.

In 2019, the three countries did just that by unilaterally increasing their mackerel quota (Henley, 2019). The EU, Norway, and the Faroe Islands responded with condemnation and the MSC suspended the certification of mackerel again. The ICES has recommended a 68% cut in current catches to restore the stock to a sustainable level by 2021, but it remains to be seen how these countries will collaboratively manage mackerel stock, especially as the abundance of mackerel continues to shift with changing climate conditions.

While climate change impacts are diverse, and a great deal of uncertainty exists, managers can research and plan for different scenarios; for example, the ICES is working with NGOs—including EDF—to compile the correct questions and push forward innovative solutions for a changing landscape. It will prove crucial to cultivate strong regional partnerships and develop large, ecosystem-based approaches (Moustahfid, 2017). Institutions should begin to incorporate uncertainty into management frameworks, and all new policies should consider climate change from the start.

China could develop solutions by promoting more research into the impact of climate change on capture fisheries and mariculture locally, regionally and globally, and the natural ecosystem services upon which these industries depend, by taking into account the potential impacts of warming waters, acidification, and altered weather, nutrient and water cycles. China may also wish to consider ways to mitigate the effects of climate change and effectively adapt to them.

7.4.2 Accelerating and Scaling Solutions Internationally

The mackerel wars example highlighted the need for research into the effects of climate change to not only understand how such effects can affect domestic fishing opportunities, but also to examine how shifting stocks could potentially create sources of conflict. Beyond producing science and information on what the future might hold, it will also be critical to develop the social and governance infrastructure needed, domestically and at international levels, to help mitigate and resolve such challenges. This infrastructure, however, does not need to be entirely focused on or driven by a need to resolve conflict. Rather, there are plenty of opportunities in the short-term to create partnerships and connectivity out of goodwill and mutual benefit on shared issues.

For example, it is no coincidence that there is such little data about fisheries in developing contexts. These countries' fisheries are often small scale, multi-species and dispersed, making centralized data collection difficult (Purcell & Pomeroy, 2015). More fundamentally, there is a severe lack of technical, financial and governance capacity in many of these fisheries. As developed nations innovate and improve their living marine resource management, these lessons should be shared and replicated in developing countries before their living marine resources fall prey to the same misfortune. Domestically, improving the technical, financial and governance capacity of developing countries' fisheries provides better food security for populations where fish is normally a vital source of protein (Hall et al., 2013). Internationally, growing our global knowledge and management of living marine resources ensures more stable markets and healthier oceans for all involved.

The Oceans and Fisheries Partnership is a shared effort between the United States Agency for International Development (USAID) and the Southeast Asian Fisheries Development Center (SEAFDEC) to use improved regional relationships to prevent and stop illegal, unreported and unregulated (IUU) fishing, and improve the Asia Pacific region's fisheries and biodiversity. The partnership uses an array of tools grounded in an ecosystems approach to fisheries management. For example, the collaboration pursues the development of catch documentation and traceability systems that help encourage transparency, identify and combat IUU and improve available data. The partnership also examines entire supply chains to leverage private sector investment to increase impacts, create new market incentives and ensure sustainable sources of funding across all its participating geographies (USAID, 2018). These large-scale collaborations are essential and benefit all parties by advancing information exchange and scaling knowledge and capacity to address some of the biggest challenges facing fisheries.

While China continues to advance its expertise and experience in management of living marine resources domestically, there are opportunities to exhibit regional and global leadership through sharing and exporting technologies, management approaches and technical know-how to other countries around the world. This is particularly relevant to the aquaculture industry, given the extensive history and amount of aquaculture operations and levels of production in China, and the increasingly strengthened systems that have been employed to monitor environmental conditions that measure yield, production and environmental impact. In addition, as China continues to advance in key areas, such as integrated and spatial management, and improved fisheries management of its domestic fisheries, many nations will likely be looking to China as a source of both inspiration and knowledge in addressing similar problems in their respective countries. Mechanisms like the Belt and Road Initiative, and related bodies such as the Belt and Road Initiative International Green Development Coalition, could serve as strong platforms to promote technical exchanges between China and many countries in the region, thereby expanding the holistic benefits of ecological civilization on a broader scale.

Chapter 8
Recommended Solutions and Ways Forward in China's Living Marine Resources Management

John Mimikakis, Hui Liu, Jacob P. Kritzer, John Virdin, Jilan Su, Ling Cao, and Fang Sun

With multiple measures and comprehensive management, China has vigorously promoted the construction of marine ecological civilization in recent years, and the state of its marine ecological environment has gradually improved (Ministry of Ecology and Environment, 2019). However, more than 1/4 of the coastal waters are still seriously polluted, 3/4 of the coastal marine ecosystems are in a sub-healthy or unhealthy state, large areas of marine habitats have been destroyed, the problems

J. Mimikakis (✉)
Environmental Defense Fund, Washington, DC, USA
e-mail: jmimikakis@gmail.com

H. Liu
Yellow Sea Fisheries Research Institute, Chinese Academy of Fishery Sciences, Qingdao, China
e-mail: liuhui@ysfri.ac.cn

J. P. Kritzer
Environmental Defense Fund, Boston, USA
e-mail: jake@neracoos.org

J. Virdin
Duke University, Durham, USA
e-mail: john.virdin@duke.edu

J. Su
Second Institute of Oceanology, Ministry of Natural Resources, Hangzhou, China
e-mail: sjl@sio.org.cn

L. Cao
Xiamen University, Xiamen, China
e-mail: caoling@xmu.edu.cn

F. Sun
Environmental Defense Fund, Beijing, China
e-mail: sunfang@cet.net.cn

H. Liu et al. (eds.), *Conservation and Sustainable Use of Living Marine Resources and Biodiversity*, Global Ocean Governance and Ecological Civilization,
https://doi.org/10.1007/978-981-19-4790-2_8

of overfishing and aquaculture overcapacity are far from being solved, and various challenges facing China's coastal and marine ecosystems still remain. In addition, the combination of pressures from the evolution of natural ecosystems, such as climate change and ocean acidification, with the aforementioned challenges will not only severely test the resilience of China's marine ecosystems, but also affect the mariculture industry and lead to cross-border migration of wild fishery populations, thereby increasing the difficulty of living marine resources management.

China has full range of laws and regulations related to the management of marine fisheries and aquaculture (e.g., the *Fisheries Law*, the *Administration of Sea Use Law*, and the *Marine Environmental Protection Law*) which are formulated in accordance with the principles of sustainable development and ecological harmony, and address the coordinated development of ecological environment protection with marine industry. Especially in recent years, China has successively issued a series of management regulations on promoting the green development of fisheries, such as the *13th Five-Year Plan for the Development of National Fisheries* (2016–2020) and *Several Opinions on Accelerating the Promotion of Green Development of Aquaculture* (2019). The Chinese government has repeatedly emphasized the following targets for fisheries development:

(1) improving quality and efficiency, reducing yields and increasing income, green development and benefiting fishermen;
(2) targeting healthy aquaculture and moderate fishing, resource protection and industry consolidation;
(3) promoting the structural reform of the fisheries supply side, and accelerating the transformation of fishery development modes;
(4) improving the standardization, greening, industrialization, organization and sustainable development of fisheries production;
(5) improving the quality, efficiency and competitiveness of fisheries development; and
(6) developing a unique path for the development of modern China fisheries characterized by high efficiency, product-safety, resource-saving and environment-friendly.

China's overall fisheries development has made significant progress during the "13th Five-Year Plan" period, and significant achievements have been made in industrial transformation and upgrading, resource conservation, equipment and technology upgrading. However, China is in a period of transition from traditional to modern fisheries, and the industry is large but not strong; there is still an incomplete fishery legal and policy system; the ecological environment of the fishery waters is fragile and difficult to protect; the fishery infrastructure is weak and the industrial efficiency urgently needs to be improved; the fisheries law enforcement cannot meet the outstanding issues of supervision (BOF, 2020). However, as the "13th Five-Year Plan" is about to come to an end, it is necessary to conduct a comprehensive and systematic assessment on the implementation of the tasks and the accomplishment of work objectives defined by the Plan. In particular, it is necessary to summarize the experience and lessons in time, in order to provide a reference for further development of fisheries during the "14th Five-Year Plan" period.

China's executive power is determined and prepared to rise to the challenge, adopting more rigorous and robust management measures and employing more advanced technical means to continuously improve its protection and sustainable development of marine biological resources. China will not only accelerate the resolution of its own issues, but will play a leading role in regional and global ocean governance through a series of comprehensive measures.

Looking ahead, China is expected to make significant progress in strengthening the protection of marine ecosystems, improving monitoring capabilities for ocean governance, developing a healthy and sustainable mariculture industry, regulating the production of marine capture fisheries, and protecting marine biodiversity. According to the latest news from the Ministry of Ecology and Environment, over the next 5–15 years, China's marine environmental quality evaluation standards will reflect more of the ecological elements of the ocean and pay special attention to typical habitats, such as mangroves, coral reefs and reed wetlands, as well as aboriginal marine fish, shore birds and other species. During the "14th Five-Year Plan" period and the near future, China expects to see improvements in offshore water quality, retention rate of natural coastlines and area of coastal wetland restoration, and an increase in "Beautiful Bays" and coastlines (housing birds and fish) that are suitable for commercial and residential use. In order to realize these fine visions, China needs to make more efforts in the following areas.

8.1 Strengthen Legal Protections for Coastal and Marine Ecosystems, While Promoting Sustainable Production

China has employed various measures, by improving quality and efficiency, returning fish farms to tidal flats, ecological redline control, zoning of aquaculture waters and treatment of aquaculture discharge, to gradually strengthen the sustainable management of its large-scale aquaculture industry and achieve a stable supply of high-quality aquatic products under the premise of ecological harmony. There is, to some extent, a disconnection between fishery-related regulations and the actual state of the industry; inoperability of certain regulations still affects the efficiency of their implementation. In order to further strengthen the management of living marine resources, it is necessary to create more powerful legal tools by taking advantage of the actual situations. China has begun to transform its huge aquaculture industry into an engine for sustainably managed safe, high-quality food production. To achieve further development, stronger legal means need to be created.

It is recommended that a new aquaculture law should be enacted, by strengthening the system design at the operational level, and to place limits on aquaculture density, waste discharge and resource (space, energy, water, and feed) use. The law should require science-based spatial limits that correspond to the environmental suitability assessment and account for environmental carrying capacity, incorporate aquaculture spatial limits into National Narine Functional Zoning, and enable the industry to optimize the value of aquaculture production while minimizing environmental impacts. Additionally, the law should establish a reporting system for aquaculture stock and inventory, as well as corresponding operation and supervision procedures, to achieve a spatial and biological basis for aquaculture permit management. This law can also make other relevant provisions for mitigating the impact of aquaculture industry on coastal and marine ecosystems, such as restricting the use of antibiotics or other chemicals.

In recent years, China has already begun to shift toward Total Allowable Catch (TAC) management in its capture fisheries. International experience shows that fishery management is more effective when output control is combined with rights-based approaches that allocate fishing quotas or the right to use fishing waters to relevant stakeholders in the fishing community and industry. Such an approach has proven more effective when rights-based approaches enable the government to address social issues, by equitably allocating access to fish among large and small industries, commercial and recreational fishing sectors and small-scale fishing communities. With the continuous improvement of fishery supervision and information collection, rights-based approaches will become increasingly practical, and fisheries law should be amended to promote their adoption.

The laws and regulations for protecting coastal wetlands and key fishery habitats should be strengthened, and conservation of living marine resources should be promoted at the institutional level. Healthy habitats are necessary for productive coastal and marine ecosystems. China should enact a strong *Marine Habitat Conservation Law* (MHCL) that strengthens protection of coastal and marine habitats and encourages significant new rehabilitation efforts that restore lost ecosystem functions and resiliency, by focusing on restoring the functions of offshore ecosystems; reserve sufficient spawning and nursery grounds for fishery organisms; and improve the ecosystem's ability to self-regulate in response to human activities and the pressure of climate change. The law should require the development of a network of marine and coastal protected areas large enough to support a biodiverse ecosystem and the production of high-value economic benefits from capture and recreational fisheries over the long term.

8.2 Implement a High-Tech Monitoring System to Combat Corruption and Illegal Fishery Operations

Over the last decade, the development of new technologies such as satellite communications, observation networks, artificial intelligence (AI), and robotics has been advancing rapidly. Innovative and disruptive technologies have continued to emerge, constantly breaking the limits of human exploration in understanding and using the ocean, and will make profound influence on the development and management of the oceans. Technological advancement will usher in a new era of exploration of ocean sciences, using satellite technology, high-resolution imaging, synthetic aperture radar (SAR), light-sensing ranging technology (Lidar) and new methane sensors to achieve more accurate and in-depth observations of the ocean. On the sea surface, wave-powered ocean robots have great potential for observing the ocean surface; the maritime drone fleet can automatically collect ocean data in various environments. Below the sea surface, unmanned ships are breaking through the limits of sea conditions, water depth and distance, and will be widely used in shipping and scientific research in the future.

China needs to promote the application of cutting-edge information technology, to strengthen and improve fishery monitoring measures, and motivate changes in the management of living marine resources. China has a long coastline, a large number of fishing boats and mariculture farms, as well as a diversified range of marine protected areas and areas controlled by the ecological redline. Since these regions and sectors are managed by different government departments, management incongruity may arise due to information asymmetry, poor communication between departments, difficulty in data sharing and regulatory loopholes. Advanced monitoring technology helps strengthen the management of marine biological resources. China's innovative capabilities in monitoring sensors, network technologies and artificial intelligence may help us build a more transparent regulatory system that operates between various agencies—and even on a global scale—to promote marine ecosystem protection, law enforcement and law abiding operations. The application of such a monitoring system on a domestic scale can expand the control of offshore fishing boats, aquatic product landing sites, mariculture farms and coastal and marine protected areas; application on a global scale can enable China to take the lead in helping other countries to ensure resource sustainability.

In addition to promoting legal operations, high-tech monitoring systems can also bring other benefits: generating large amounts of new data, greatly improving China's awareness of the health of its coastal and marine ecosystems; enabling the government to deal with pollution accidents and other emergencies in real time, so as to protect public health and food safety; and helping other countries to understand the impact, while encouraging their cooperation with China in exploring ways to mitigate these impacts.

8.3 Develop Mariculture Sustainably Under the Framework of Ecosystems

Presently, mariculture has become a crucial part of China's coastal and inshore marine ecosystems, similar to many other coastal countries globally. Faced with this indisputable fact, it is necessary to employ ecosystem thinking when evaluating and managing the industry. To develop and expand the mariculture industry, we need to adhere to the coordinated development of ecology, society, economy and cultural values; and incorporation of mariculture into ecosystem governance is necessary to ensure its sustainable development. A comprehensive assessment of the biome and its ecological features, level of productivity and carrying capacity, as well as the impact of aquaculture on the natural marine ecosystems is necessary, for scientifically and reasonably determining the optimum aquaculture species, density and mode of operation, as a basis for developing new modes of ecosystem-based mariculture. From the perspective of the fisheries authorities, the above mentioned assessment results can be and should be attached to the management of aquaculture permits.

It is necessary to promote the comprehensive governance of mariculture and its coordinated development with various marine ecosystem services. United Nations Sustainable Development Goal (SDG) #14 should be fully embodied in China's mariculture development plan, that is:

(1) to protect and sustainably use the oceans and marine resources to promote sustainable development;
(2) to strengthen the supervision and control of the space, water, feed and other resources utilized by mariculture;
(3) to improve quality and efficiency in all aspects;
(4) to promote the concept of "reduce, recycle, reuse" in mariculture and its related industrial chains, and guide the research and development of sustainable seafood production technologies with science and technology programs;
(5) to pay attention to the constraints of water, energy and space resources, and possibly climate change, and conduct input–output accounting for unit space area, energy consumption and water consumption; and
(6) to use economic levers to encourage circular economy in mariculture, and boost the enthusiasm and self-consciousness of enterprises to participate in ecosystem governance.

8.4 Upgrade Technologies Continuously, to Improve Quality and Efficiency of Mariculture

8.4.1 Strengthen Quality Assurance and Prevent Aquatic Product Safety Risks

High-quality, healthy and safe feed formulation and production should become minimum requirements for mariculture industry in China. To achieve this goal, legislation should be strengthened to strictly prohibit the use of trash fish as feed or feed ingredients. This is also to protect China's fishery resources and mariculture water environment, reduce the occurrence of mariculture diseases, and improve food safety, ecological safety and international competitiveness of China mariculture products. A comprehensive aquatic product quality and safety control mechanism should be established, with a framework of indicators and technology system for aquatic product quality and safety evaluation and risk assessment, consisting the full breadth of quality and safety assurance technologies, products, standards and specifications, for the entire aquatic food production chain in China.

China needs investment on basic research and technology development on mariculture environmental pollution monitoring, supervision and early warning. This should involve strengthening the monitoring, evaluation and management of the illegal use and discharge (to natural water bodies) of antibiotics and other chemicals in mariculture. To facilitate this effort, relevant basic, applied and comprehensive research should be advanced, focusing on "baseline survey" on antibiotics and other pollution sources in and contamination of mariculture and fishery waters. In addition, basic research should be coupled with the establishment of a corresponding monitoring, evaluation and early warning technology system. Finally, there will also be a need to strengthen and stabilize the financial support for research and development of mariculture ecosystem conservation technologies, to reinforce the systematic control of environmental pollution in mariculture operations, and to establish a comprehensive, systematic and efficient management institution.

8.4.2 Develop Smart, Low-Carbon Technologies and Promote Green Transformation of the Mariculture Industry

Using a "brain" of artificial intelligence, and deploying intelligent perception, transmission and information processing, the mariculture Internet of Things (IOT) may be used to establish a computer-based monitoring, decision making and management platform for intensive mariculture farms, to help reduce operational risks and improve the automation of operation. This can lead to the establishment of a digital and intelligent mariculture network platform for the entire process of mariculture management, including water quality monitoring, automatic feeding, disease prevention, water circulation and purification control, and logistics supervision. China has vigorously promoted the Internet Plus initiative in agriculture, which would promote agricultural modernization through a networked, intelligent and accurate "planting, breeding and processing" mode of ecological agriculture. In this context, the transition to intelligent technology should be accelerated in mariculture to improve production methods, and inject vitality into the development of high-density, intensive and large-scale mariculture modes.

Agricultural preference funds should be used to guide the transformation and upgrade of mariculture production methods and the development of low-carbon technologies, and promote the treatment of wastewater from fish farms, the application of high-quality feeds with high conversion rates, and the application of facilities and equipment for precision-feeding and emission reduction. Water resource use fees should be gradually implemented to promote water and energy conserving and ecological harmonious mariculture. A number of more conservation-oriented mariculture modes are already available and should be further demonstrated and promoted, including land-based industrial recirculating aquaculture systems (RAS), integrated multi-trophic aquaculture (IMTA) in ponds and shallow seas, and ecological sea ranching. At the same time, China should rationally plan for the mid-term and long-term layout of the mariculture industry; vigorously develop high-efficiency and low-carbon mariculture demonstration parks with successful wastewater treatments; strengthen integrated technology demonstrations, and set up cutting edge mariculture systems with high ecological and economic benefits. For developing technology-intensive engineering mariculture, such as deep-sea net cages and mariculture factory vessels, it is important to strengthen scientific investment on research and development of key technical links and hotspot technologies.

8.5 Restore Marine Ecosystem Functions to Improve Resilience to Global Change

China has already taken important steps to conserve coastal and marine ecosystems through the redlining process. However, more should be done to restore lost habitats, including mangroves, seagrass beds, tidal marshes and flats, and coral reefs. These habitats provide a valuable array of ecosystem services; among them, providing nursery and spawning grounds for a diverse array of marine organisms, filtering and detoxifying pollutants, protecting the coast from erosion and buffering the effects of climate change. It is not enough to merely protect existing coastal and sea areas. If the country's coastal and marine ecosystems are to withstand the impacts of pollution and climate change, and continue to be a source of tremendous prosperity and food production, China should consider the following steps to undertaking a large scale effort, guided by ecological science, to restore lost ecosystem functions and services:

8.5.1 Establish a National Marine Ecological Report Card on the Health of China's Coastal and Marine Ecosystems

China's economic goals for marine capture fisheries, mariculture, marine tourism and other industries should be based on a sound ecological understanding of what the country's coastal and marine ecosystems can support. To develop a strong scientific foundation, China should consider establishing a comprehensive ecological assessment (i.e., a marine ecological report card) on the health of the country's marine and estuarine zones. Being different from the *Bulletin on the State of China's Marine Environment* issued by the former State Oceanic Administration and the *Bulletin on the State of China's Marine Ecological Environment* issued by the Ministry of Ecology and Environment, this report card should assess the cumulative effects of the intense uses of China's living marine resources by fisheries and mariculture, and the impacts of pollution, development, tourism industries and climate change, on the health of China's coastal and marine ecosystems. It should evaluate the integrity of key ecosystem functions and services, such as water and nutrient cycles and critical habitats, and measure their resiliency to climate change and other future pressures. The report should also recommend ways that China can improve the capacity and resilience of coastal and marine ecosystems to support seafood production and tourism and conserve biodiversity. The report should be made publicly available and regularly updated.

8.5.2 Develop a National Plan of Action to Restore Lost Marine Ecosystem Functions and Services

Like most other countries, China's oceans, coasts and estuaries are often under the jurisdiction of different administrative departments, and smooth communication and coordination between departments usually becomes an issue. China should consider formulating a comprehensive management plan similar to the Chesapeake Bay Program, described in Sect. 7.2.2, in order to coordinate the conservation actions of different departments. This plan would promote coordination and the division of labor between departments, jointly maintain sound ecosystem services and functions and optimize China's long-term economic benefits. The plan should involve the Ministry of Agriculture and Rural Affairs, the Ministry of Ecology and Environment, the Ministry of Natural Resources, and relevant provincial and municipal administrative departments in coastal areas. The plan should also provide guidelines for various governance actions, such as setting fishing quotas, implementing zoning management of mariculture, using ecological redlines to protect habitats, restoring coastal and marine habitats, controlling pollution and allocating functional areas through marine spatial planning. In view of the special geographical, economic and social status of the Bohai Sea as China's inland sea, it is recommended that the Bohai Sea shall be used as a pilot area for the national plan of action to restore lost marine ecosystems.

8.5.3 Assess and Mitigate the Impact of Climate Change on Marine Living Resources

Climate change is already having a wide variety of significant effects on living marine resources around the world, which threaten to become worse over time. China could develop solutions to this issue by promoting more research into the impact of climate change on the country's capture fisheries and mariculture, and the natural ecosystem services upon which these industries depend. It is also important to evaluate how climate change will likely affect fisheries and food production across the entire Asia Pacific region and around the world. Assessments should take into account the potential impacts of warming waters, acidification, and altered weather, nutrient and water cycles.

In addition, China may wish to consider ways to not only mitigate the effects of climate change, but effectively adapt to it. For example, scientists could examine how provincial governments could work together to manage fisheries, how the industry could remain profitable as species shift in their ranges, how research programs could be designed for breeding heat- or acid-tolerant mariculture species, and how international fish sharing agreements could be developed or strengthened.

In short, China needs to undertake multiple approaches and coordinated research, to support fisheries production, biodiversity conservation, and to enhance resilience to resource development, environmental pollution and climate change.

8.6 Maritime Silk Road Networking to Promote Sustainable Ocean Governance

There is but one global ocean, and people all over the world are affected by the health of its ecosystems. The Maritime Silk Road Initiative, the sea-orientated component of the Belt and Road Initiative, represents a historic opportunity for China to demonstrate leadership in global marine governance and advance the UN SDGs. Under the Silk Road Initiative, China should consider creating a network of partnerships with countries from Asia, Africa and Europe to encourage mutual learning and promote joint actions that promote a healthy ocean and sustainable ocean governance.

China's leadership is important, as many developing countries face similar challenges in managing their living marine resources, but they typically lack China's scientific and technical expertise, governance capacity and financial resources. As a result, the marine resources in these countries are among the most poorly managed and threatened in the world. China can promote sustainability along the Maritime Silk Road by promoting information sharing and collaborations that build educational, scientific and technical capacity in partner countries. Key topics could include managing marine resources sustainably, promoting economic development, improving food security for vulnerable peoples, combatting illegal fishing and building the capacity of women in fishing communities and supply chains.

China could also demonstrate leadership by using the Maritime Silk Road Initiative to catalyze the development of regional and global approaches that can mitigate the impact of climate change on living marine resources. In Asia and Africa, climate change could lead to greater conflict among nations as wild fish migrate across borders, and changes in the productivity of mariculture and capture fisheries strain the ability of nations to produce food. The Maritime Silk Road Initiative could provide the region with the leadership and institutional platform to develop urgently needed, collaborative solutions to address the effects of climate change.

8.7 Carry Out a National Plan for Implementation of the UN 2030 Agenda for Sustainable Development Goals

China should continue to fulfill its commitments and strictly implement the country's plan for the United Nations 2030 SDGs, and promote marine ecosystem-based management by: strengthening the protection of major typical ecosystems; scientifically drawing a redline on marine ecosystem protection; improving the network of marine reserves; establishing a nationwide, real-time, online monitoring system for marine environment; and conducting research on marine ecological compensation and indemnity mechanisms.

To effectively regulate capture fisheries, China should improve the capability for conservation and management of living marine resources, implement a science-based management plans for fishery resources, strictly control fishing intensity, continue to implement a fishing moratorium system, and sustainably use existing fishery resources.

To remove unreasonable fisheries subsidies, China has already set goals and policy measures on fishing boat controls during the "13th Five-Year Plan" period, and amended the *Provisions on the Administration of Fishery Licensing*. There is also a pressing need to keep the intensity of crackdowns on TNFVs, and rapidly reduce fuel subsidies. Targeted support should be given to reduce the number of fishing boats, fishermen's career conversion, maintenance and renovation of fishing ports, and standardization of aquaculture ponds etc. China should also actively participate in World Trade Organization fisheries subsidies negotiation.

To increase the economic benefits of sustainable use of marine resources for Small Island Developing States (SIDS), China will need to play an active role through the BRI initiative, helping strengthen governance, providing aquaculture technology support such as energy saving, emissions reduction and recirculating aquaculture systems (RAS). China will also need to promote South-South cooperation on sustainable management of fisheries and tourism.

8.8 Continuously Promoting Gender Equity in Fisheries Sector

Currently, about 59.51 million people are engaged (on a full-time, part-time or occasional basis) in capture fisheries (39.0 million people) and aquaculture (20.5 million people) globally, and women account for 14% of the total, with shares of 19% in aquaculture and 12% in capture fisheries (FAO, 2020). Although they play a vital role in the seafood sector, women in many countries around the world face challenges of inequitable access to opportunity, minimal representation and unequal benefits received from their participation.

In addition to achieving fundamental human principles of social justice and fairness, there is a growing body of evidence that gender equality can lead to improved household income, productivity and nutritional security (Hillenbrand et al., 2015). Addressing the issue of gender inequality is a critical component of sustainable development globally, so much so that it is promoted by United Nations SDG #5: Achieve gender equality and empower all women and girls. Although, aquaculture farms and processing factories in China help close the gender gap by employing more women as temporary workers, inequality remains in terms of their lower wages and less employment opportunities. There is a strong need in China to build understanding of gender roles in management of living marine resource and the gender aspects of alternative policies, in order to build greater gender equity. Additional research and efforts are also needed to improve women's education, as well as social and economic opportunities and responsibilities in fisheries and aquaculture in China.

8.9 Epilogue

The ocean is a cornerstone of planetary health. Its sustainable management is central to ensuring the long-term viability of marine ecosystems, economies, and communities. The challenge of sustainably managing the ocean's living marine resources, in particular, is a global imperative, and few nations are as pivotal to the solution as China. With its vast marine economy, unique biodiversity, and outsized role in global seafood production and consumption, China is uniquely positioned to lead in protecting and restoring marine ecosystems.

As the pressures of climate change, overfishing, and coastal development mount, China's marine ecosystems have reached a tipping point. The nation's ambitious efforts to address these challenges through policy reforms, scientific innovation, and stronger enforcement reflect its recognition of the stakes and the political will to address the strain on its marine resources.

For example, China has instituted strict seasonal fishing closures, some of the most extensive in the world, to allow depleted fish stocks to recover. Efforts to establish marine protected areas and restore critical habitats are beginning to help rebuild ecosystems that have suffered from decades of overexploitation and coastal development. The government has also introduced mariculture acreage limits and shut down illegal operations to curb environmental degradation and keep aquaculture development within ecological limits.

At the same time, technological innovation is becoming a key driver of change. Investments in monitoring systems are beginning to improve compliance, while sustainable mariculture frameworks are being advanced to minimize environmental impacts and enhance resilience to climate change.

Despite these achievements, more must be done. Legal protections for living marine resources must be strengthened, habitat restoration accelerated, and monitoring systems further expanded. Addressing these issues will require not only further domestic reforms but also robust regional and global cooperation to promote sustainability with partner governments and regional institutions, especially as climate change reshapes the dynamics of marine ecosystems worldwide.

China's commitment to building an ecological civilization offers an opportunity to redefine what sustainable ocean management looks like. By scaling these efforts, China can secure its own food security, protect livelihoods, and demonstrate leadership in fostering a future where marine ecosystems thrive alongside human development. The path forward is challenging but holds immense promise—for China, the region, and the world.

Acknowledgements We would like to thank the China Council for International Cooperation on Environment and Development (CCICED) of the Ministry of Ecology and Environment of People's Republic of China (MEE) for convening and funding the special policy study on Global Ocean Governance and Ecological Civilization. We appreciate the helpful and constructive comments from Dr. Su Jilan, Dr. Jan-gunnar Winther, Dr. Art Hansen and Dr. Liu Shijin, which greatly improved this report. We would give special thanks to Mr. Li Yonghong, Mr. Zhang Huiyong, Ms. Yao Ying, Mr. Li Panwen and Ms. Fei Chengbo from CCICED, and Ms. Wang Minfang from the Second Institute of Oceanology, Ministry of Natural Resources, for your tireless work to ensure the success of this project. We also would like to thank Dr. Lisa Svensson from the United Nations Environment, Mr. Nishan Degranian from the World Economic Forum, Dr. Dai Minhan from Xiamen University, Dr. Wang Juying from the National Marine Environmental Monitoring Centre of MEE, Dr. Sun Song from the Institute of Oceanology of Chinese Academy of Sciences, Dr. Han Baoxin from the South China Institute of Environmental Sciences of MEE and Dr. Birgit Njaastad from Norwegian Polar Institute, for your valuable support of this work.

References

Abbott, J. K., Lloyd-Smith, P., Willard, D., & Adamowicz, W. (2018). Status-quo management of marine recreational fisheries undermines angler welfare. *Proceedings of the National Academy of Sciences, 115*(36), 8948–8953. https://doi.org/10.1073/pnas.1809549115

Agar, J. J., Stephen, J. A., Strelcheck, A., & Diagne, A. (2014a). The Gulf of Mexico red snapper IFQ program: The first five years. *Marine Resource Economics, 29*(2), 177–198. https://doi.org/10.1086/676825

Agar, J. J., Stephen, J. A., Strelcheck, A., et al. (2014b). The Gulf of Mexico red snapper IFQ program: The first five years. *Marine Resource Economics, 29*(2), 177–198.

Agriculture and Agri-Food Canada (AAFC). (2017). *Sector trend analysis: fish and seafood trends in China.* Global analysis report, prepared by Mengchao Chen. Ottawa, ON, Canada. Retrieved from http://www.agr.gc.ca/resources/prod/Internet-Internet/MISB-DGSIM/ATS-SEA/PDF/6869-eng.pdf

An, S., Li, H., Guan, B., Zhou, C., Wang, Z., Deng, Z., Zhi, Y., Liu, Y., Xu, C., Fang, S., Jiang, J., Hong, L., & Li, H. (2007). China's natural wetlands: Past problems, current status, and future challenges. *Ambio, 34*, 335–342.

Arias-Ortiz, A., Serrano, O., Masqué, P., Lavery, P. S., Mueller, U., Kendrick, G. A., et al. (2018). A marine heatwave drives massive losses from the world's largest seagrass carbon stocks. *Nature Climate Change, 8*, 338. https://doi.org/10.1038/s41558-018-0096-y

Arnold, S. N., Steneck, R. S., & Mumby, P. J. (2010). Running the gauntlet: Inhibitory effects of algal turfs on the processes of coral recruitment. *Marine Ecology Progress Series, 414*, 91–105. https://doi.org/10.3354/meps08724

Bakun, A. (1990). Coastal ocean upwelling. *Science, 247*, 198–201. https://doi.org/10.1126/science.247.4939.198

Bakun, A. (2017). Climate change and ocean deoxygenation within intensified surface-driven upwelling circulations. *Philosophical Transactions of the Royal Society A: Mathematical, Physical and Engineering Sciences, 375*, 20160327. https://doi.org/10.1098/rsta.2016.0327

Berkeley, S. A., Hixon, M. A., Larson, R. J., & Love, M. S. (2011). Fisheries sustainability via protection of age and spatial distribution of fish populations. *Fisheries, 29*(8), 23–32. https://doi.org/10.1577/1548-8446(2004)29[23:FSVPOA]2.0.CO;2

Blomeyer, R., Sanz, A., Stobberup, K., Goulding, I., & Pauly, D. (2012). The role of China in world fisheries. *Directorate-General for Internal Policies, Policy Department B: Structural and Cohesion Policies—Fisheries.* For the European Parliament's Committee on Fisheries. Brussels.

Boenish, R., Willard, D., Kritzer, J. P., & Reardon, K. (2020). Fisheries monitoring: Perspectives from the United States. *Aquaculture and Fisheries, 5*, 131–138.

Briggs, J. C. (1974). *Marine zoogeography.* McGraw-Hill.

H. Liu et al. (eds.), *Conservation and Sustainable Use of Living Marine Resources and Biodiversity*, Global Ocean Governance and Ecological Civilization,
https://doi.org/10.1007/978-981-19-4790-2

British Sea Fishing. (2019). *The Mackerel war. British sea fishing*. Retrieved from https://britishseafishing.co.uk/the-mackerel-wars/#:~:text=The%20Mackerel%20War%20is%20an,Islands%20which%20began%20in%202010

Burden, M., Kleisner, K., Landman, J., Priddle, E., & Ryan, K. (eds.). (2017). *Climate-related impacts on fisheries management and governance in the North East Atlantic*. Workshop report. Retrieved from https://www.edf.org/sites/default/files/documents/climate-impacts-fisheries-NE-Atlantic_0.pdf

Bureau of Fisheries (BOF), Ministry of Agriculture and Rural Affairs. (1949–2020). *China fisheries statistical yearbook*. China Agriculture Press. (农业农村部渔业局 (1949–2020). 中国渔业统计年鉴. 农业出版社, 北京. In Chinese).

Bureau of Fisheries (BOF), Ministry of Agriculture and Rural Affairs. (2020). "Thirteenth Five-Year Plan" Fisheries Highlights Serial | Accelerating the pace of transformation and upgrading, high-quality development of fisheries has achieved substantial results (农业农村部渔业局, 2020. "十三五" 渔业亮点连载 | 转型升级步伐加快 渔业高质量发展取得实效. In Chinese). Retrieved December 26, 2020, from http://www.moa.gov.cn/xw/bmdt/202012/t20201208_6357773.htm

Burgess, M., McDermott, G., Owashi, B., Peavey Reeves, L., Clavelle, T., Ovando, D., Wallace, B., Lewison, R., Gaines, S., & Costello, C. (2018). Protecting marine mammals, turtles, and birds by rebuilding global fisheries. *Science, 359*(6381), 1255–1258. https://doi.org/10.1126/science.aao4248

Cao, W., & Wong, M. (2007). Current status of coastal zone issues and management in China: A review. *Environment International, 33*(7), 985–992.

Cao, W., Hong, H., Yue, S., Ding, Y., & Zhang, Y. (2003). Nutrient loss from an agricultural catchment and landscape modeling in southeast China. *Bulletin of Environmental Contamination and Toxicology, 71*(4), 761–767.

Cao, L., Naylor, R., Henriksson, P., Leadbitter, D., Metian, M., Troell, M., & Zhang, W. (2015). China's aquaculture and the world's wild fisheries. *Science, 347*, 133–135.

Cao, L., Chen, Y., Dong, S., Hanson, A., Huang, B., Leadbitter, D., Naylor, R., et al. (2017). Opportunity for marine fisheries reform in China. *Proceedings of the National Academy of Sciences, 114*(3), 435–442. https://doi.org/10.1073/pnas.1616583114

Carugati, L., Gatto, B., Rastelli, E., Lo Martire, M., Coral, C., Greco, S., & Danovaro, R. (2018). Impact of mangrove forests degradation on biodiversity and ecosystem functioning. *Scientific Reports, 8*. https://doi.org/10.1038/s41598-018-31683-0

Casteñeda, A., Maaz, J., Requeña, N., & Chan, S. (2011). Managed access in Belize. In *Proceedings of the 64th Gulf and Caribbean Fisheries Institute*. Puerto Morelos, Mexico.

CBP. (1987, December 15). *The 1987 Chesapeake Bay Agreement*. Retrieved from https://www.chesapeakebay.net/content/publications/cbp_12510.pdf

CBP. (2000, June 28). *Chesapeake 2020*. Retrieved from https://www.chesapeakebay.net/content/publications/cbp_12081.pdf

CBP. *Population growth*. Chesapeake Bay Program. Retrieved January 23, 2012, from https://www.chesapeakebay.net/issues/population_growth

CBP. (2019). *Chesapeake Bay Watershed Agreement*. Retrieved from https://www.chesapeakebay.net/what/what_guides_us/watershed_agreement

CCICED Task Force. (2013). *Ecosystem issues and policy options addressing the sustainable development of China's ocean and coasts* [Report] (p. 493). China Environmental Science Press. (国合会课题组. 2013. 中国海洋可持续发展的生态环境问题与政策研究. 北京:中国环境出版社, 493).

Chen, Z. B., & Chen, W. (2006). Reflections on the turbot incident. *Ocean and Fisheries, 2006*(12), 10 (陈智兵, 陈文, 2006. 多宝鱼事件引起的反思。海洋与渔业, 2006(12), 10 In Chinese with English abstract).

Chen, Z., Qiu, Y., & Xu, S. (2011). Changes in trophic flows and ecosystem properties of the Beibu Gulf ecosystem before and after the collapse of fish stocks. *Ocean & Coastal Management, 54*(8), 601–611.

Chesapeake Bay Foundation. (2018). *State of the bay 2019* [Report]. Retrieved from https://www.cbf.org/about-the-bay/state-of-the-bay-report/

Cheung, W. W. L., Jones, M. C., Lam, V. W. Y., Miller, D., Ota, Y., Teh, L., & Sumaila, U. R. (2017). Transform high seas management to build climate resilience in marine seafood supply. *Fish and Fisheries, 18*, 254–263. https://doi.org/10.1111/faf.12177

China Association for Science and Technology (CAST), China Fisheries Society (CFS). (2020). *2018–2019 Fishery science development report*. China Science and Technology Press. (中国科学技术协会、中国水产学会, 2020。2018–2019 水产学学科发展报告。中国科学技术出版社, 2020. In Chinese).

Choi, F., Gouhier, T., Lima, F., Rilov, G., Seabra, R., & Helmuth, B. (2019). Mapping physiology: Biophysical mechanisms define scales of climate change impacts. *Conservation Physiology, 7*. https://doi.org/10.1093/conphys/coz028

Chopin, T., Buschmann, A. H., Halling, C., Troell, M., Kautsky, N., Neori, A., Kraemer, G. P., Zertuche-González, J. A., Yarish, C., & Neefus, C. (2001). Integrating seaweeds into marine aquaculture systems: A key toward sustainability. *Journal of Phycology, 37*, 975–986. https://doi.org/10.1046/j.1529-8817.2001.01137.x

Christensen, V. (1996). Managing fisheries involving predator and prey species. *Reviews in Fish Biology and Fisheries, 6*, 417–442. Retrieved from: https://link.springer.com/article/10.1007/BF00164324

Clark, C. W. (2006). Fisheries economics: Why is it so widely misunderstood? *Population Ecology, 48*, 95–98.

Clavelle, T., Lester, S. E., Gentry, R., & Froehlich, H. E. (2019). Interactions and management for the future of marine aquaculture and capture fisheries. *Fish and Fisheries, 20*, 368–388. https://doi.org/10.1111/faf.12351

Costello, C., Gaines, S. D., & Lynham, J. (2008). Can catch shares prevent fisheries collapse? *Science, 321*(5896), 1678–1681. https://doi.org/10.1126/science.1159478

Costello, C., Ovando, D., Clavelle, T., Strauss, C. K., Hilborn, R., Melnychuk, M. C., & Leland, A. (2016). Global fishery prospects under contrasting management regimes. *Proceedings of the National Academy of Sciences, 113*(18), 5125–5129. https://doi.org/10.1073/pnas.1520420113

Costello, C., Ovando, D., Hilborn, R., Gaines, S., Deschenes, O., & Lester, S. (2012). Status and solutions for the world's unassessed fisheries. *Science, 338*(6106), 517–520.

Craik, W. (1992). The Great Barrier Reef Marine Park: Its establishment, development, and current status. *Marine Pollution Bulletin, 25*(5–8), 122–133. https://doi.org/10.1016/0025-326X(92)90215-R

Davidson, K., Gowen, R. J., Harrison, P. J., Fleming, L. E., Hoagland, P., & Moschonas, G. (2014). Anthropogenic nutrients and harmful algae in coastal waters. *Journal of Environmental Management, 146*, 206–216. https://doi.org/10.1016/j.jenvman.2014.07.002

De Groot, R., Brander, L., Ploeg, S. V., Costanza, R., Bernard, F., Braat, L., & Beukering, P. V. (2012). Global estimates of the value of ecosystems and their services in monetary units. *Ecosystem Services, 1*(1), 50–61. https://doi.org/10.1016/j.ecoser.2012.07.005

De Silva, S., & Soto, D. (2009). Climate change and aquaculture: Potential impacts, adaptation and mitigation. In K. Cochrane, C. De Young, D. Soto, & T. Bahri (Eds.), *Climate change implications for fisheries and aquaculture: overview of current scientific knowledge* (FAO Fisheries and Aquaculture Technical Paper No. 530). Food and Agriculture Organization of the United Nations.

Diana, J. S. (2009). Aquaculture production and biodiversity conservation. *BioScience, 59*(1), 27–38. https://doi.org/10.1525/bio.2009.59.1.7

Diaz, R. J., & Rosenberg, R. (2008). Spreading dead zones and consequences for marine ecosystems. *Science, 321*, 926–929. https://doi.org/10.1126/science.1156401

Drumm, A., Echeverria, J., & Almendarez, M. (2011). *Sustainable finance strategy and plan for the belize protected area system*. Retrieved from http://protectedareas.gov.bz/download/392/

Du, J., Ye, G., Chen, B., & Zheng, X. (2014). Changes in the marine trophic index of Chinese marine area. *Biodiversity Science, 22*(04), 532–539. (杜建国, 叶观琼, 陈彬, 郑新庆 (2014). 中国海

域海洋生物的营养级指数变化特征。生物多样性 22(04), 532–539. In Chinese with English abstract).

Duarte, C. M., Wu, J., Xiao, X., Bruhn, A., & Krause-Jensen, D. (2017). Can seaweed farming play a role in climate change mitigation and adaptation. *Frontiers in Marine Science, 4*. https://doi.org/10.3389/fmars.2017.00100

Duarte, B., Martins, I., Rosa, R., Matos, A. R., Roleda, M. Y., Reusch, T. B. H., Engelen, A. H., Serrão, E. A., Pearson, G. A., Marques, J. C., Caçador, I., Duarte, C. M., & Jueterbock, A. (2018). Climate change impacts on seagrass meadows and macroalgal forests: An integrative perspective on acclimation and adaptation potential. *Frontiers in Marine Science, 5*(190), 1–23. https://doi.org/10.3389/fmars.2018.00190

Economic Impacts. (n.d.). *Chesapeake Bay Today*. Retrieved from https://sites.google.com/a/cornell.edu/chesapeake-bay-today/economic-impacts

Endean, R. (1982). Crown-of-thorns starfish on the great barrier reef. *Endeavour, 6*(1), 10–14. https://doi.org/10.1016/0160-9327(82)90004-7

Eng, C. T., Paw, J. N., & Guarin, F. Y. (1989). The environmental impact of aquaculture and the effects of pollution on coastal aquaculture development in Southeast Asia. *Marine Pollution Bulletin, 20*(7), 335–343. https://doi.org/10.1016/0025-326X(89)90157-4

Evans, L. S., Ban, N. C., Schoon, M., & Nenadovic, M. (2014). Keeping the 'great' in the Great Barrier Reef: Large-scale governance of the Great Barrier Reef Marine Park. *International Journal of the Commons, 8*(2), 396. https://doi.org/10.18352/ijc.405

Fabinyi, M., & Liu, N. (2014). The Chinese policy and governance context for global fisheries. *Ocean and Coastal Management, 96*, 198–202.

FAO. (2014). *The state of world fisheries and aquaculture 2012* (p. 223). Rome.

FAO. (2016). *The state of world fisheries and aquaculture 2014* (p. 200). Rome.

FAO. (2019). FAO yearbook. *Fishery and Aquaculture Statistics 2017/FAO annuaire. Statistiques des pêches et de l'aquaculture 2017/FAO anuario. Estadísticas de pesca y acuicultura 2017*, p. 108.

FAO (2020). *The state of world fisheries and aquaculture 2020*. Rome. https://doi.org/10.4060/ca9229en

FAO, IFAD, UNICEF, WFP, WHO. (2020). In brief to the state of food security and nutrition in the world 2020. In *Transforming food systems for affordable healthy diets*. FAO. https://doi.org/10.4060/ca9699en

Fernandes, L., Day, J., Lewis, A., Slegers, S., Kerrigan, B., Breen, D., & Stapleton, K. (2005). Establishing representative no-take areas in the Great Barrier Reef: Large-scale implementation of theory on marine protected areas. *Conservation Biology, 19*(6), 1733–1744. https://doi.org/10.1111/j.1523-1739.2005.00302.x

Figgatt, M., Hyde, J., Dziewulski, D., Wiegert, E., Kishbaugh, S., Zelin, G., & Wilson, L. (2017). Harmful algal bloom-associated illnesses in humans and dogs identified through a pilot surveillance system—New York, 2015. MMWR. *Morbidity and Mortality Weekly Report, 66*, 1182–1184. https://doi.org/10.15585/mmwr.mm6643a5

Final Rule Traceability. (2018). Retrieved July 12, 2019, from https://www.iuufishing.noaa.gov/RecommendationsandActions/RECOMMENDATION1415/FinalRuleTraceability.aspx

Foley, J. R. (2012). Managed access: moving towards collaborative fisheries sustainability in Belize. In *Proceedings of the 12th International Coral Reef Symposium*. Cairns, Australia.

Froehlich, H. E., Gentry, R. R., & Halpern, B. S. (2018). Global change in marine aquaculture production potential under climate change. *Nature Ecology and Evolution, 2*, 1745–1750. https://doi.org/10.1038/s41559-018-0669-1

Froehlich, H. E., Afflerbach, J. C., Frazier, M., & Halpern, B. S. (2019). Blue growth potential to mitigate climate change through seaweed offsetting. *Current Biology, 29*, 3087-3093.e3. https://doi.org/10.1016/j.cub.2019.07.041

Fu, X. M., Guang, H. S., Wang, Y. N., Shao, C. L., Wang, C. Y., Li, G. Q., Liu, G. X., Sun, S. C., Zeng, X. Q., Ye, Z. J., & Guang, H. S. (2009). Investigation on the status of mangrove resources and medicinal research in China II. Resources status, protection and management. *Journal of*

Ocean University of China, 39(4), 705–711. (傅秀梅,王亚楠, 邵长伦, 王长云, 李国强, 刘光兴, 孙世春, 曾晓起, 叶振江, 管华诗, (2009). "中国红树林资源状况及其药用研究调查 II. 资源现状,保护与管理." 中国海洋大学学报(自然科学版) *39*(004), 705–711. In Chinese with English abstract).

Fujita, R., et al. (2019). Assessing and managing small-scale fisheries in Belize. In S. Salas, M. Barragán-Paladines, & R. Chuenpagdee (Eds.), *Viability and sustainability of small-scale fisheries in Latin America and The Caribbean*. MARE Publication Series (Vol. 19). Springer.

Fujita, R., Epstein, L., Battista, W., Karr, K., Higgins, P., Landman, J., & Carcamo, R. (2017). Scaling territorial use rights in fisheries (TURFs) in Belize. *Bulletin of Marine Science, 93*(1), 137–153.

Gaines, S. D., Costello, C., Owashi, B., Mangin, T., Bone, J., Molinois, J. G., Burden, M., Dennis, H., Halpern, B. S., Kappel, C. V., Kleisner, K. M., & Ovando, D. (2018). Improved fisheries management could offset many negative effects of climate change. *Science Advances, 4*(8), 8.

Gao, Y., Fu, J., Zeng, L., Li, A., Li, H., Zhu, N., Liu, R., Liu, A., Wang, Y., & Jiang, G. (2014). Occurrence and fate of perfluoroalkyl substances in marine sediments from the Chinese Bohai Sea, Yellow Sea, and East China Sea. *Environmental Pollution, 194*, 60–68.

Garcia, S. M., Kolding, J., Rice, J., Rochet, M. J., Zhou, S., Arimoto, T., & Smith, A. D. M. (2012). Reconsidering the consequences of selective fisheries. *Science, 335*, 1045–1047.

Gentry, R. R., Froehlich, H. E., Grimm, D., Kareiva, P., Parke, M., Rust, M., Gaines, S. D., & Halpern, B. S. (2017). Mapping the global potential for marine aquaculture. *Nature Ecology & Evolution, 1*, 1317–1324. https://doi.org/10.1038/s41559-017-0257-9

Gershman, D., Wondolleck, J., & Yafee, S. (2012). *Great Barrier Reef marine park*. Marine Ecosystem-Based Management in Practice. Retrieved from http://webservices.itcs.umich.edu/drupal/mebm/?q=node/56

Gilman, E. L., Ellison, J., Duke, N. C., & Field, C. (2008). Threats to mangroves from climate change and adaptation options: A review. *Aquatic Botany, 89*, 237–250. https://doi.org/10.1016/j.aquabot.2007.12.009

Godfrey, M. (2018). *China's seafood imports surged in 2017, while its export growth continues to slow*. SeafoodSource. Retrieved from https://www.seafoodsource.com/news/supply-trade/chinas-seafood-imports-surged-in-2017-while-its-export-growth-continued-to-slow

Golden, C. D., Allison, E. H., Cheung, W. W., Dey, M. M., Halpern, B. S., McCauley, D. J., Smith, M., Vaitla, B., Zeller, D., & Myers, S. S. (2016). Nutrition: Fall in fish catch threatens human health. *Nature, 534*(7607), 317–320. https://doi.org/10.1038/534317a

Government Accountability Office (GAO). (2016). *Additional actions could advance efforts to incorporate climate information into management decisions* (Report No. GAO-16-827). US Government Accountability Office.

Granéli, E., & Turner, J. T. (2006). Ecology of harmful algae: 15 tables. In *Ecological studies*. Springer, Berlin.

Great Barrier Reef Marine Park Zoning Plan 2003. (2003). Retrieved from http://www.gbrmpa.gov.au/__data/assets/pdf_file/0015/3390/GBRMPA-zoning-plan-2003.pdf

Grech, A., Pressey, R. L., & Day, J. C. (2015). Coal, cumulative impacts, and the Great Barrier Reef. *Conservation Letters, 9*(3), 200–207. https://doi.org/10.1111/conl.12208

Grieve, B. D., Hare, J. A., & Saba, V. S. (2017). Projecting the effects of climate change on Calanus finmarchicus distribution within the U.S. Northeast Continental Shelf. *Scientific Reports, 7*, 6264.

Griffin, J. (2019). Why tiny Belize is a world leader in protecting the ocean. *The Guardian*. Retrieved from https://www.theguardian.com/environment/2019/aug/14/why-tiny-belize-is-a-world-leader-in-ocean-protection

Gulf Wild. (2019). Retrieved July 12, 2019, from https://gulfwild.com/faqs.php

Gutierrez, M., Daniels, A., Jobbins, G., Almazor, G. G., & Montenegro, C. (2020). *China's distant-water fishing fleet: Scale, impact, and governance*. ODI.

Hall, O., Holby, O., Kollberg, S., & Samuelsson, M. O. (1992). Chemical fluxes and mass balances in a marine fish cage farm. IV. Nitrogen. *Marine Ecology Progress Series, 89*(1), 81–91.

Hall, S. J., Hilborn, R., Andrew, N. L., & Allison, E. H. (2013). Innovations in capture fisheries are an imperative for nutrition security in the developing world. *Proceedings of the National Academy of Sciences, 110*(21), 8393–8398. https://doi.org/10.1073/pnas.1208067110

Halpern, B. S., Walbridge, S., Selkoe, K. A., Kappel, C. V., Micheli, F., D'Agrosa, C., & Watson, R. (2008). A global map of human impact on marine ecosystems. *Science, 319*, 948–952.

Han, Y. (2018). Marine fishery resources management and policy adjustment in China since 1949. *Chinese Rural Economy*, (9), 14–28. (韩杨. 2018. 1949 年以来中国海洋渔业资源治理与政策调整. 中国农村经济, (9), 14–28. In Chinese with English abstract).

He, Y., & Chen, L. (2020). Outdated cognition and institution: Double-lagged marine ecological compensation and its improvement. *Cross-Strait Legal Science, 22*(1), 37–46. (何跃军， 陈淋淋(2020). "认知滞后与制度滞后: 海洋生态补偿的双重滞后与改进." 海峡法学 22(01), 37–46. In Chinese with English abstract).

He, P., Xu, S., Zhang, H., Wen, S., Dai, Y., Lin, S., & Yarish, C. (2008). Bioremediation efficiency in the removal of dissolved inorganic nutrients by the red seaweed, *Porphyra yezoensis*, cultivated in the open sea. *Water Research, 42*, 1281–1289. https://doi.org/10.1016/j.watres.2007.09.023

He, Q., Bertness, M. D., Bruno, J. F., Li, B., Chen, G., Coverdale, T. C., Altieri, A. H., Bai, J., Sun, T., Pennings, S. C., Liu, J., Ehrlick, P. R., & Cui, B. (2014). Economic development and coastal ecosystem change in China. *Scientific Reports, 4*(1), 5995.

Henley, J. (2019). Iceland accused of putting mackerel stocks at risk by increasing its catch. *The Guardian*. Retrieved from https://www.theguardian.com/environment/2019/nov/21/iceland-accused-of-putting-mackerel-stocks-at-risk-by-increasing-its-catch

Hillenbrand, E., Karim, N., Mohanraj, P., & Wu, D. (2015). *Measuring gender-transformative change: a review of literature and promising practices*. CARE USA. Working Paper.

Hou, J. X., & Feng, X. M. (2010). Economic governance for sustainable utilization of marine fishing resource. *Shandong Economy, 159*(4), 48–54. (侯景新, 冯小妹. 2010. 海洋渔业资源可持续利用的经济管理. 山东经济, *159*(4), 48–54. In Chinese).

Hsieh, C., Yamauhi, A., Nakazawa, T., & Wang, W. (2009). Fishing effects on age and spatial structures undermine population stability of fishes. *Aquatic Sciences, 72*, 165–178. Retrieved from https://link.springer.com/article/10.1007%2Fs00027-009-0122-2

Huang, Z. G. (2008). *Marine species and their distribution in China's seas* (Lin, J., [Translation based on the first edition published in 1994] 2001) (p. 1191). Ocean Press. (黄宗国, 2008. 中国海洋生物种类与分布 (增订版)。北京: 海洋出版社, pp. 1191 In Chinese).

Huang, S., & He, Y. (2019). Management of China's capture fisheries: Review and prospect. *Aquaculture and Fisheries, 4*, 173–182.

Huang, Z. G., & Lin, M. (2012). *Species diversity in China seas* (Vols. 1–2). Ocean Press. (黄宗国和林茂, 2012. 中国海洋物种多样性 (上 下册)。北京: 海洋出版社 In Chinese).

Huang, S. L., & Tang, Y. (2019). Review and prospect of theories of fisheries management and China's practice. *Journal of Fisheries of China, 43*(1), 211–231 (黄硕琳, 唐议. 2019. 渔业管理理论与中国实践的回顾与展望. 水产学报, *43*(1), 211–231. In Chinese with English abstract).

Hughes, T. P., Huang, H. U. I., & Young, M. A. L. (2013). The wicked problem of China's disappearing coral reefs. *Conservation Biology, 27*(2), 261–269.

Hughes, T. P., Anderson, K. D., Connolly, S. R., et al. (2018). Spatial and temporal patterns of mass bleaching of corals in the Anthropocene. *Science, 359*, 80–83.

Hutchings, J. A., & Reynolds, J. D. (2004). Marine fish population collapses: Consequences for recovery and extinction risk. *BioScience, 54*(4), 297–309. https://doi.org/10.1641/0006-3568(2004)054[0297:MFPCCF]2.0.CO;2

Igulu, M. M., Nagelkerken, I., Dorenbosch, M., Grol, M. G. G., Harborne, A. R., Kimirei, I. A., Mumby, P. J., Olds, A. D., & Mgaya, Y. D. (2014). Mangrove habitat use by juvenile reef fish: Meta-analysis reveals that tidal regime matters more than biogeographic region. *PLoS One, 9*, e114715. https://doi.org/10.1371/journal.pone.0114715

Jensen, F., Frost, H., Thøgersen, T., Andersen, P., & Andersen, J. L. (2015). Game theory and fish wars: The case of the Northeast Atlantic mackerel fishery. *Fisheries Research, 172*, 7–16. https://doi.org/10.1016/j.fishres.2015.06.022

Jia, X. P., Chen, J. C., Chen, H. G., Qi, Z. H., Zhu, C. B., Yang, J., Meng, S. L., Song, C., & Shen, X. Q. (2017). Aquaculture environmental assessment and governance. In Q. Tang (Ed.), *Environment-friendly mariculture development strategy: New ideas, new tasks, new approaches.* (pp. 14–34). Science Press. (贾晓平, 陈家长, 陈海刚, 齐占会, 朱长波, 杨健, 孟顺龙, 宋超, 新强. 2017. 水产养殖环境评估与治理. [唐启升等主编] 环境友好型水产养殖发展战略: 新思路、新任务、新途径. 北京: 科学出版社, 268–310. In Chinese).

Johnson, C. L., Runge, J. A., Curtis, K. A., Durbin, E. G., Hare, J. A., Incze, L. S., Link, J. S., Melvin, G. D., O'Brien, T. D., & Van Guelpen, L. (2011). Biodiversity and ecosystem function in the Gulf of Maine: Pattern and role of zooplankton and pelagic nekton. *PLoS One, 6*, e16491. https://doi.org/10.1371/journal.pone.0016491

Jordà, G., Marbà, N., & Duarte, C. M. (2012). Mediterranean seagrass vulnerable to regional climate warming. *Nature Climate Change, 2*, 821–824. Retrieved from https://imedea.uib-csic.es/master/cambioglobal/Modulo_III_cod101608/tema%204-temperatura/arti%CC%81culos%20tema%204/Jord%C3%A0%20et%20al%202012%20(Nature%20Climate%20Change).pdf

Kenchington, R., & Day, J. (2011). Zoning, a fundamental cornerstone of effective marine spatial planning: Lessons learnt from the Great Barrier Reef, Australia. *Journal of Coastal Conservation, 15*(2), 271–278. https://doi.org/10.1007/s11852-011-0147-2

Knol, M. (2010). Scientific advice in integrated ocean management: The process towards the Barents Sea plan. *Marine Policy, 34*(2), 252–260. https://doi.org/10.1016/j.marpol.2009.07.009

Kritzer, J. P. (2020). Influences of at-sea fishery monitoring on science, management, and fleet dynamics. *Aquaculture and Fisheries, 5*, 107–112.

Lavoie, D., Denman, K. L., & MacDonald, R. W. (2010). Effects of future climate change on primary productivity and export fluxes in the Beaufort Sea. *Journal of Geophysical Research*, 115. https://doi.org/10.1029/2009JC005493

Lester, S. E., Stevens, J. M., Gentry, R. R., Kappel, C. V., Bell, T. W., Costello, C. J., & White, C. (2018). Marine spatial planning makes room for offshore aquaculture in crowded coastal waters. *Nature Communications, 9*(1). https://doi.org/10.1038/s41467-018-03249-1

Liang, C., Xian, W., & Pauly, D. (2018). Impacts of ocean warming on China's fisheries catches: An application of "mean temperature of the catch" concept. *Frontiers in Marine Science, 5*, 1–7. https://doi.org/10.3389/fmars.2018.00026

Lin, B., Boenish, R., Kritzer, J. P, Jiang, Y., Wang, S., & Liu, M. (2021). Reproductive dynamics of a swimming crab (*Monomia haanii*) in the world's crab basket. *Fisheries Research*, 236. https://doi.org/10.1016/j.fishres.2020.105828

Liu, G. (2014). Food losses and food waste in China: A first estimate. *OECD Food, Agriculture and Fisheries Papers* (p. 66). OECD Publishing.

Liu, H. (2016). National aquaculture law and policy: China. In N. Bankes, I. Dahl, & D. L. VanderZwaag (Eds.), *Aquaculture law and policy—Global, regional and national perspectives* (pp. 238–265). Edward Elgar Publishing.

Liu, H., & Su, J. L. (2017). Vulnerability of China's nearshore ecosystems under intensive mariculture development. *Environmental Science and Pollution Research, 24*, 8957–8966.

Liu, D., Keesing, J. K., He, P., Wang, Z., Shi, Y., & Wang, Y. (2013). The world's largest macroalgal bloom in the Yellow Sea, China: Formation and implications. *Estuarine, Coastal and Shelf Science, 129*, 2–10.

Liu, H. B., Sun, L., Qi, J. T., & Yang, F. L. (2007). Comparative study on marine fisheries management policies between China and South Korea. *Pacific Journal*, (12), 69–77. (刘洪滨, 孙丽, 齐俊婷, 杨凤丽. 2007. 中韩两国海洋渔业管理政策的比较研究. 太平洋学报, (12), 69–77. In Chinese with English abstract).

Liu, H., Huang, X. P., Wang, Y. L., Liang, Z. R., Gu, B., Su, J. L. (2016). Newly discovered seagrass bed and its ecological characteristics in the coastal area of Caofeidian, Bohai Sea. *Chinese Journal of Ecology, 35*(7), 1677–1683 (刘慧, 黄小平, 王元磊, 梁洲瑞, 古彬, 苏纪兰, 2016. 渤海曹妃甸新现的海草床及其生态特征。生态学杂志, *35*(7), 1677–1683 In Chinese with English abstract).

Liu, H., Sun, L., Wang, J., Wang, Q., & Tang, Q. (2017). Current status, problems and countermeasures of environmental-friendly mariculture. In Q. Tang (Ed.), *Environment-friendly mariculture development strategy: New ideas, new tasks, new approaches* (pp. 14–34). Science Press. (刘慧, 孙龙启, 王建坤, 王清印, 唐启升, 2017. 环境友好型水产养殖现状、问题与应对建议. [唐启升等主编] 环境友好型水产养殖发展战略: 新思路、新任务、新途径. 北京:科学出版社, 14–34. In Chinese).

Liu, H., Jiang, Z. J., Yu, L. J., Xuan, J. L., Shang, W. T., et al. (2021). *Spatial management for marine aquaculture*. Science Press. (刘慧, 蒋增杰, 于良巨, 宣基亮, 尚伟涛等 (著), 2021. 海水养殖空间管理。科学出版社, 北京。In Chinese).

Liu, J. Y. (2013). Status of marine biodiversity of the China seas. *PLoS One, 8*(1), e50719–e50819.

Liu, L., Wang, H., & Yue, Q. (2020). China's coastal wetlands: Ecological challenges, restoration, and management suggestions. *Regional Studies in Marine Science, 37*, 101337.

Liu, R., Cui, Y., Xu, F., & Tang, Z. (1983). Ecology of macrobenthos of the East China Sea and adjacent waters. In *Paper presented at the Proceeding of International Symposium on the Sedimentation on the Continental Shelf, with special reference to the East China Sea.* China Ocean Press.

Liu, X., Wang, Y., Costanza, R., Kubiszewski, I., Xu, N., Yuan, M., & Geng, R. (2019). The value of China's coastal wetlands and seawalls for storm protection. *Ecosystem Services, 36*, 100905.

Liu, Z., Cui, B., & He, Q. (2016). Shifting paradigms in coastal restoration: Six decades' lessons from China. *Science of the Total Environment, 566–567*, 205–214.

Lowman, D. M., Fisher, R., Holliday, M. C., McTee, S. A., & Stebbins, S. (2013). *Fisheries monitoring roadmap: A guide to evaluate, design and implement an effective fishery monitoring program that incorporates electronic monitoring and electronic reporting tools.* Environmental Defense Fund. Retrieved from https://www.edf.org/sites/default/files/FisheryMonitoringRoadmap_FINAL.pdf

Lu, X. (2005). *Study on sustainable utilization and effective management of marine fishery resources in China.* Huazhong Agricultural University. (卢秀容 (2005). 中国海洋渔业资源可持续利用和有效管理研究。华中农业大学硕士学位论文. In Chinese with English abstract).

Luo, J. X., & Xu, S. W. (2017). Analysis of protein intake of Chinese people. *Prospect of Agriculture, 10*, 71–76 (罗洁霞, 许世卫. 2017. 中国居民蛋白质摄入量状况分析. 农业展望, *10*, 71–76. In Chinese).

Luo, M. (2019). *Remote sensing monitoring and ecosystem service value assessment of coastal salt marshes in China in 2015.* Zhejiang University. (罗敏 (2019). 2015 年中国海岸带盐沼遥感监测与生态服务价值评估。浙江大学硕士学位论文. In Chinese with English abstract).

Ma, Z., Melville, D. S., Liu, J., Chen, Y., Yang, H., Ren, W., Zhang, Z., Piersma, T., & Li, B. (2014). Rethinking China's new great wall. *Science, 346*(6212), 912–914.

Mai, K. S. (2020). History and prospect of aquaculture nutrition and aquafeed industry of China. *Feed Industry, 41*(1), 2–6 (麦康森, 2020. 中国水产动物营养研究与饲料工业的发展历程与展望。饲料工业, *41*(1), 2–6. In Chinese with English abstract).

Mallory, T. G. (2013). China's distant water fishing industry: Evolving policies and implications. *Marine Policy, 38*, 99–108.

Marine Stewardship Council. (2016). *Mackerel wins back its certified sustainable status* [Press release]. Retrieved from https://www.msc.org/media-centre/press-releases/mackerel-wins-back-its-certified-sustainable-status

Martinez, V. A., Castañeda, A., Gongora, M., & Wade, B. (2018). Managed *access: A rights-based approach to managing small scale fisheries in Belize. Tenure and user rights in fisheries 2018.* Achieving sustainable development goals by 2030 [PowerPoint slides]. Retrieved from http://www.fao.org/3/CA2430EN/ca2430en.pdf

McDonald, G., Carcamo, R., Fujita, R., Gedamke, T., Karr, K., & Wilson, J. (2014). A multi-indicator framework for adaptive management of data-limited fisheries with a case study from Belize. In *Proceedings of the 67th Gulf and Caribbean Fisheries Institute.*

McElderry, H. (2013). *Electronic monitoring in the shore-side hake fishery 2004–2010.* PFMC EM Workshop Agenda (Item B.1).

McField, M., Kramer, P., Alvarez Filip, L., Drysdale, I., Rueda Flores, M., Giro Petersen, A., & Soto, M. (2018) *Report card for the Mesoamerican Reef*. Healthy Reefs Initiative. Retrieved from http://www.healthyreefs.org

McKenzie, J. (2019). Will GOB comply with Belize network of NGOs and pass new fisheries bill. *Love News*. Retrieved from http://lovefm.com/business-companies-organizations/will-gob-comply-with-belize-network-of-ngos-and-pass-new-fisheries-bill/

Ministry of Agriculture (MOA), Ministry of Environmental Protection (MEP). (2015). Report on the State of the Fishery Eco-Environment in China, 2015. (农业部, 环境保护部, 2015. 2015 年中国渔业生态环境状况公报)。Retrieved July 26, 2020, from https://www.cafs.ac.cn/info/1387/24114.htm

Ministry of Agriculture and Rural Affairs (MARA), Ministry of Ecology and Environment. (2018). Report on the State of the Fishery Eco-Environment in China, 2018. (农业农村部, 生态环境部, 2018. 2018 年中国渔业生态环境状况公报。). Retrieved July 26, 2020, from https://www.cafs.ac.cn/info/1387/33583.htm

Ministry of Agriculture (MOA). (2017). *National Sea Ranch Demonstration Zone Construction Plan (2017–2025)*. Retrieved December 26, 2020, from http://www.moa.gov.cn/nybgb/2017/201711/201802/t20180201_6136235.htm. (农业部, 2017. 国家级海洋牧场示范区建设规划 (2017–2025). In Chinese).

Ministry of Agriculture (MOA). (2018). *Ministry of Agriculture: China's aquatic product import and export trade reached a new high in 2017, and it is expected that the trade surplus will narrow down in 2018*. (农业部: 2017 年我国水产品进出口贸易再创新高 预计 2018 全年贸易顺差将收窄). Retrieved June 2, 2018, from http://www.moa.gov.cn/xw/bmdt/201803/t20180314_6138388.htm

Ministry of Agriculture and Rural Affairs (MARA), Ministry of Ecology and Environment (MEE). (2018). *2018 Report on the state of the fishery eco-environment in China,* Retrieved July 26, 2020, from https://www.cafs.ac.cn/info/1387/33583.htm (农业农村部, 生态环境部, 2018. 2018 年中国渔业生态环境状况公报. In Chinese).

Ministry of Agriculture and Rural Affairs (MARA). (2019). List of drugs and other compounds prohibited for use in food animals (draft for comments). Retrieved December 26, 2020, from http://www.moa.gov.cn/xw/bmdt/201909/t20190903_6327159.htm. (农业农村部, 2019. 食品动物禁止使用的药物及其他化合物清单 (征求意见稿)。In Chinese).

Ministry of Ecology and Environment (MAA). (2020). *The Ministry of Ecology and Environment announces the national ecological environment quality profile in 2019*. Retrieved January 3, 2021, from http://www.mee.gov.cn/xxgk2018/xxgk/xxgk15/202005/t20200507_777895.html. (生态环境部, 2020. 生态环境部公布2019年全国生态环境质量简况。In Chinese).

Ministry of Finance, Ministry of Agriculture. (2015). *Notice about adjusting domestic fisheries and aquaculture fuel subsidy policy in order to promote sustainable and healthy fisheries development* (Document 2015 No. 499). Retrieved from http://jjs.mof.gov.cn/zhengwuxinxi/zhengcefagui/201507/t20150709_1272152.html

Ministry of Foreign Affairs of the People's Republic of China (MFA). (2017). *China's progress report on implementation of the 2030 Agenda for Sustainable Development*. Retrieved from https://www.fmprc.gov.cn/web/ziliao_674904/zt_674979/dnzt_674981/qtzt/2030kcxfzyc_686343/P020170824650025885740.pdf

Ministry of Natural Resources (MNR). (2020). *China Ocean Economy Statistics 2019*. Retrieved July 26, 2020 from http://gi.mnr.gov.cn/202005/t20200509_2511614.html (自然资源部. 2020. 2019 年中国海洋经济统计公报. In Chinese).

Morrison, S. (2016). Major fish tale: Red snapper in the Gulf of Mexico. *Natural Resources Environment, 31*(1), 13–17.

Moustahfid, H. (2017). *Current actions, identified solutions and opportunities in addressing the effects of climate change on fisheries and aquaculture* [Conference presentation]. United Nations ICP-18 Meeting, May 15–19, 2017.

Murdiyarso, D., Purbopuspito, J., Kauffman, J. B., Warren, M. W., Sasmito, S. D., Donato, D. C., Manuri, S., Krisnawati, H., Taberima, S., & Kurnianto, S. (2015). The potential of Indonesian mangrove forests for global climate change mitigation. *Nature Climate Change, 5*, 1089–1092. https://doi.org/10.1038/nclimate2734

Murray, A. G., & Peeler, E. J. (2005). A framework for understanding the potential for emerging diseases in aquaculture. *Preventive Veterinary Medicine, 67*(2–3), 223–235. https://doi.org/10.1016/j.prevetmed.2004.10.012

National Central Station for Fisheries Technology Promotion (NCSFTP). (2020). *Monitoring Report on China Recreational Fishery Development*. Retrieved December 30, 2020, from https://mp.weixin.qq.com/s/3YAOdBeixvOfOQ6xA-2KkQ. (全国水产技术推广总站, 2020. 中国休闲渔业发展监测报告 (2020)。In Chinese).

Neori, A., Chopin, T., Troell, M., Buschmann, A. H., Kraemer, G. P., Halling, C., Shpigel, M., & Yarish, C. (2004). Integrated aquaculture: Rationale, evolution, and state of the art emphasizing seaweed biofiltration in modern mariculture. *Aquaculture, 231*(1–4), 361–391. https://doi.org/10.1016/j.aquaculture.2003.11.015

Nikolek, G., de Jong, B., & Pan, C. (2018). China's changing tides: Shifting consumption and trade position of Chinese seafood. *Rabobank RaboResearch*. Retrieved from https://research.rabobank.com/far/en/sectors/animal-protein/chinas_changing_tides.html

Niu, Z. G., Gong, P., Cheng, X., Guo, J. H., Wang, L., Huang, H. B., Shen, S. Q., Wu, Y. Z., Wang, X. F., Wang, X. W., Ying, Q., Liang, L., Zhang, L. N., Wang, L., Yao, Q., Yang, Z. Z., Guo, Z. Q., & Dai, Y. J. (2009). Geographical characteristics of China's wetlands derived from remotely sensed data. *Science in China Series D - Earth Sciences, 52*(6), 723–738.

NOAA Fisheries. (2013). History of management of Gulf of Mexico red snapper. *NOAA Fisheries*. Retrieved August 24, 2019, from https://www.fisheries.noaa.gov/history-management-gulf-mexico-red-snapper

NOAA Fisheries. (2017). Commercial fisheries annual landings index. *NOAA Fisheries*. Retrieved August 24, 2019, from https://foss.nmfs.noaa.gov/apexfoss/f?p=215:200:9135774540998

NOAA Fisheries. (2017). Gulf of Mexico reef fish historical amendments and rulemaking (1983–2017). *NOAA Fisheries*. Retrieved July 8, 2019, from https://www.fisheries.noaa.gov/action/gulf-mexico-reef-fish-historical-amendments-and-rulemaking-1983-2017

Ocean and Fisheries Bureau of Zhejiang Province. (2018). *Targets achieved for pilot of quota control on catch fishery. China Fisheries Daily*. (浙江省海洋与渔业局. 2018.《中国渔业报》专访黄志平局长: "浙江试点"完成预定目标。[登录时间: 2020年2月19日] In Chinese). Retrieved February 19, 2020, from http://www.oceanol.com/zhuanti/201808/14/c80057.html

Olsen, E., Gjøsæter, H., Røttingen, I., Dommansnes, A., Fossum, P., & Sandberg, P. (2007). The Norwegian ecosystem-based management plan for the Barents Sea. *ICES Journal of Marine Science, 64*(4), 599–602. https://doi.org/10.1093/icesjms/fsm005

Organisation for Economic Co-operation and Development, Food and Agricultural Organization (OECD-FAO). (2017). *OECD-FAO Agricultural Outlook, OECD Agriculture statistics* (database). Retrieved from http://www.economyworld.net:9091/economyworld/detail/init?infoId=319094

Ottersen, G., Olsen, E., van der Meeren, G. I., Dommasnes, A., & Loeng, H. (2011). The Norwegian plan for integrated ecosystem-based management of the marine environment in the Norwegian Sea. *Marine Policy, 35*(3), 389–398. https://doi.org/10.1016/j.marpol.2010.10.017

Pang, S. J., Liu, F., Shan, T. F., Xu, N., Zhang, Z. H., Gao, S. Q., Chopin, T., & Sun, S. (2010). Tracking the algal origin of the Ulva bloom in the Yellow Sea by a combination of molecular, morphological and physiological analyses. *Marine Environmental Research, 69*(4), 207–215.

Paulson Institute. (2016). *Blueprint of coastal wetland conservation and management in China*. Paulson Institute. Retrieved from https://www.paulsoninstitute.org/conservation/wetlands-conservation/blueprint-of-coastal-wetland-conservation-and-management-in-china/

Pauly, D., Christensen, V., Dalsgaard, J., Froese, F., & Torres, F. J. (1998). Fishing down marine food webs. *Science, 279*(5352), 860–863. Retrieved from https://science.sciencemag.org/content/279/5352/860.full

Pauly, D., & Palomares, M. L. (2004). Fishing down marine food web: It is far more pervasive than we thought. *Bulletin of Marine Science, 76*(2), 197–212. Retrieved from https://www.ingentaconnect.com/content/umrsmas/bullmar/2005/00000076/00000002/art00003

Pauly, D., & Zeller, D. (2015). *Sea around us*. Retrieved from http://seaaroundus.org

Perry, C. T., Alvarez-Filip, L., Graham, N. A. J., Mumby, P. J., Wilson, S. K., Kench, P. S., Manzello, D. P., Morgan, K. M., Slangen, A. B. A., Thomson, D. P., Januchowski-Hartley, F., Smithers, S. G., Steneck, R. S., Carlton, R., Edinger, E. N., Enochs, I. C., Estrada-Saldívar, N., Haywood, M. D. E., Kolodziej, G., … MacDonald, C. (2018). Loss of coral reef growth capacity to track future increases in sea level. *Nature, 558*, 396–400. https://doi.org/10.1038/s41586-018-0194-z

Pettersen, E. (2015). *Integrated marine management: Norway's methodology and experience* [PowerPoint slides]. Norwegian Environmental Agency. Retrieved from http://www.varam.gov.lv/in_site/tools/download.php?file=files/text/Finansu_instrumenti/EEZ_2009_2014/7_10_2015_semin_pr/2_Integrated_Marine_Management_Plans_Eirik_Drablos_Pettersen.pdf

Phoenix News Media (凤凰网). (2018). Retrieved August 8, 2019, from http://news.ifeng.com/a/20181108/60150230_0.shtml

Pinsky, M. L., Worm, B., Fogarty, M. J., Sarmiento, J. L., & Levin, S. A. (2013). Marine taxa track local climate velocities. *Science, 341*, 1239–1242. https://doi.org/10.1126/science.1239352

Powers, S. P., & Anson, K. (2016). Estimating recreational effort in the Gulf of Mexico red snapper fishery using boat ramp cameras: Reduction in federal season length does not proportionally reduce catch. *North American Journal of Fisheries Management, 36*(5), 1156–1166. https://doi.org/10.1080/02755947.2016.1198284

Purcell, S., & Pomeroy, R. (2015). Driving small-scale fisheries in developing countries. *Frontiers in Marine Science, 2*(44). https://doi.org/10.3389/fmars.2015.00044

Qi, Z. H., Shi, R. J., Yu, Z. H., Han, T. T., Li, C. H., Xu, S. M., Xu, S. N., Liang, Q. Y., Yu, W., Lin, H. Z., & Huang, H. H. (2019). Nutrient release from fish cage aquaculture and mitigation strategies in Daya Bay, southern China. *Marine Pollution Bulletin, 146*, 399–407.

Qiu, J. N. (2011). China faces up to 'terrible' state of its ecosystems. *Nature, 471*(7336), 19.

Ren, A. J, Yang, Z. Y., Dai Y. J., et al. (2012). Research on aquatic product demand forecasting in China. *Journal of Shanghai Ocean University, 21*(1), 145–150. (任爱景, 杨正勇, 戴亚娟等. 2012. 我国水产品需求预测研究. 上海海洋大学学报, 21(1), 145–150. In Chinese with English abstract).

Rice, J., & Gislason, H. (1996). Patterns of change in the size spectrum of numbers and diversity of the North Sea fish assemblage, as reflected in surveys and models. *ICES Journal of Marine Science, 53*, 1214–1225.

Roheim, C. (2004). Trade liberalization in fish products: Impacts on sustainability of international markets and fish resources. In A. Aksoy, & J. Beghin (Eds.), *Global agricultural trade and developing countries*. The World Bank.

Roxy, M. K., Modi, A., Murtugudde, R., Valsala, V., Panickal, S., Prasanna Kumar, S., Ravichandran, M., Vichi, M., & Lévy, M. (2016). A reduction in marine primary productivity driven by rapid warming over the tropical Indian Ocean. *Geophysical Research Letters, 43*, 826–833. https://doi.org/10.1002/2015GL066979

Schive, P. (2018). *Ecosystem approach—Norwegian marine integrated management plans* [PowerPoint slides]. Norwegian Ministry of Climate and Environment.

Schram, F. R. (2010). Checklist of marine biota of china seas. *Journal of Crustacean Biology, 30*(2), 339–339.

Science News. (2019). *There comes again the green tide Ulva prolifera!* (科学网, 2019. 浒苔又来了 !In Chinese). Retrieved August 8, 2019, from http://news.sciencenet.cn/sbhtmlnews/2019/6/347132.shtm?id=347132

Shen, G., & Heino, M. (2014). An overview of marine fisheries management in China. *Marine Policy, 44*, 265–272. https://doi.org/10.1016/j.marpol.2013.09.012

Silva, S. S. D., & Soto, D. (2014). *Climate change and aquaculture: potential impacts, adaptation and mitigation* (FAO Technical Documents No. 1–81).

South Atlantic Fish Management Council (SAFMC). (2019). *Fishery ecosystem plan*. Retrieved from https://safmc.net/habitat-and-ecosystems/fishery-ecosystem-plan/

Southeast Data, Assessment, and Review (SEDAR). (2013). *SEDAR 31 Gulf of Mexico red snapper stock assessment report*. Retrieved from http://sedarweb.org/docs/sar/SEDAR%2031%20SAR-%20Gulf%20Red%20Snapper_sizereduced.pdf

Srinivasan, U. T., Watson, R., & Sumaila, R. (2012). Global fisheries losses at the exclusive economic zone level, 1950 to present. *Marine Policy, 36*, 544–549.

Staples, K. W., Chen, Y., Townsend, D. W., & Brady, D. C. (2019). Spatiotemporal variability in the phenology of the initial intra-annual molt of American lobster (Homarus americanus Milne Edwards, 1837) and its relationship with bottom temperatures in a changing Gulf of Maine. *Fisheries Oceanography, 28*(4), 268–485. https://doi.org/10.1111/fog.12425

State Oceanic Administration, 2011–2018. Bulletin of China's Marine Environment. (2010–2017). (国家海洋局, 2011–2018. 中国海洋环境状况公报 (2010–2017) In Chinese).

State Oceanic Administration. (2018). *National Marine Inspectorate's feedback to Jiangsu Province on the special inspects on land reclamation*. Retrieved January 16, 2018, from http://www.soa.gov.cn/xw/hyyw_90/201801/t20180114_59954.html (国家海洋局网站. 国家海洋督察组向江苏反馈围填海专项督察情况. In Chinese).

Steneck, R. S., Arnold, S. N., Boenish, R., de León, R., Mumby, P. J., Rasher, D. B., & Wilson, M. W. (2019). Managing recovery resilience in coral reefs against climate-induced bleaching and hurricanes: A 15 year case study from Bonaire, Dutch Caribbean. *Frontiers in Marine Science, 6*. https://doi.org/10.3389/fmars.2019.00265

Sui, X. R., Zhan, X. F., & Fang, J. W. (2017). Analysis on China aquatic food demand. *Modern Agriculture Science and Technology, 21*, 293–295. (隋昕融, 詹夏菲, 樊佳伟. 2017. 我国水产品消费市场预测分析. 现代农业科技, *21*, 293–295. In Chinese).

Sun, C., Zhong, K., Ge, R., Zhao, Y., Liu, X., & Lin, T. (2016). Landscape pattern changes of coastal wetland in Nansha District of Guangzhou City in recent 20 years. In F. Bian, & Y. Xie (Eds.), *Geo-informatics in resource management and sustainable ecosystem* (pp. 408–416). Springer.

Sun, C., Wang, W. W., & Liang, G. F. (2017). The analysis of market equilibrium in demand and supply of aquatic products in China. *China Fishery Economy, 35*(2), 4–11. (孙琛，王威巍，梁鸽峰.2017. 中国水产品市场供求平衡分析.中国渔业经济, *35*(2), 4–11. In Chinese).

Sun, Q. W., Liu, H., Shang, W. T., Yu, L. J., Jiang, X. P., You, J. Y., Strand, Ø. (2020). Spatial planning of aquaculture in Sanggou Bay and surrounding sea areas. Submitted manuscript.

Sun, X., Wu, M. Q., Xing, Q. G., Song, X. D., Zhao, D. H., Han, Q. Q., & Zhang, G. Z. (2018). Spatio-temporal patterns of *Ulva prolifera* blooms and the corresponding influence on chlorophyll-a concentration in the southern Yellow Sea, China. *Science of the Total Environment, 640–641*, 807–820.

Sun, Z. G., Sun, W. G., Tong, C., Zeng, C. S., Yu, X., & Mou, X. J. (2015). China's coastal wetlands: Conservation history, implementation efforts, existing issues and strategies for future improvement. *Environment International, 79*, 25–41.

Szuwalski, C. S., Burgess, M. G., Costello, C., & Gaines, S. D. (2016). High fishery catches through trophic cascades in China. *Proceedings of the National Academy of Sciences, 114*(4), 717–721. https://doi.org/10.1073/pnas.1612722114

Tang, D., Di, B., Wei, G., Oh, I., & Wang, S. (2006). Spatial, seasonal and species variations of harmful algal blooms in the South Yellow Sea and East China Sea. *Hydrobiologia, 568*, 245–253. https://doi.org/10.1007/s10750-006-0108-1

Tang, Q. S. (ed.). (2012). *Regional oceanography of China seas. Fisheries oceanography* (p. 405). Ocean Press (唐启升[主编]. 2012. 中国区域海洋学—渔业海洋学. 北京: 海洋出版社, pp. 450. In Chinese).

Tang, Q. S. (ed.). (2013). *Strategic study on sustainable development of China aquaculture* (p. 339). China Agriculture Press. (唐启升[主编], 2013. 中国养殖业可持续发展战略研究—水产养殖卷。中国农业出版社, 北京, pp. 408. In Chinese).

Tang, Q. S. (ed.). (2017). *Environment-friendly mariculture development strategy: New ideas, new tasks, new approaches* (p. 460). Science Press. (唐启升 主编. 2017. 环境友好型水产养殖发展战略: 新思路、新任务、新途径. 北京: 科学出版社, pp. 460. In Chinese).

Tang, Q. S. (2019). The development and position of fisheries stock enhancement, sea ranching and enhancement fisheries. *China Fishery*, (5), 28–29. (唐启升. 2019. 渔业资源增殖、海洋牧场、增殖渔业及其发展定位. 中国水产, (5), 28–29. In Chinese with English abstract).

Tang, Y., & Zou, W. H. (2010). Assessment of the legal system for China fisheries resource conservation and management. *Resource Science, 32*(1), 28–34. (唐议, 邹伟红. 2010. 中国渔业资源养护与管理的法律制度评析. 资源科学, *32*(1), 28–34. In Chinese).

The Great Barrier Reef Marine Park Authority. (2002). *Technical information sheet #5: Representative areas program background and history*. Retrieved from http://www.gbrmpa.gov.au/__data/assets/pdf_file/0010/6211/tech_sheet_05.pdf

The San Pedro Sun. (2017). *2017–2018 Queen conch fishing quota is set*. Retrieved from https://www.sanpedrosun.com/business-and-economy/2017/11/04/2017-2018-queen-conch-fishing-quota-set/

The State Council Leading Group Office of Poverty Alleviation and Development. (2020). Retrieved December 25, 2020, from http://www.cpad.gov.cn/ (国务院扶贫开发领导小组办公室, 2020。http://www.cpad.gov.cn/ 2020 年12月25日登录。).

Tian, B., Wu, W., Yang, Z., & Zhou, Y. (2016). Drivers, trends, and potential impacts of long-term coastal reclamation in China from 1985 to 2010. *Estuarine, Coastal and Shelf Science, 170*, 83–90.

United States Agency for International Development (USAID). (2018). *USAID oceans and fisheries partnership fact sheet*. Retrieved from https://www.usaid.gov/asia-regional/fact-sheets/usaid-oceans-and-fisheries-partnership

United States National Oceanic and Atmospheric Administration (NOAA). (n.d.). *History of management of Gulf of Mexico red snapper*. Retrieved from https://www.fisheries.noaa.gov/history-management-gulf-mexico-red-snapper

United States National Oceanic and Atmospheric Administration (NOAA). (2017). *Status of stocks 2017: Annual report to Congress on the status of US fisheries*. Washington, DC.

United States National Oceanic and Atmospheric Administration (NOAA). (n.d.). *History of management of Gulf of Mexico red snapper*. Retrieved from https://www.fisheries.noaa.gov/history-management-gulf-mexico-red-snapper

United States Drug Administration Natural Resoruces Conservation Service (USDA NRCS). (2008). *Chesapeake Bay Watershed Program* [Map]. Retrieved from https://www.nrcs.usda.gov/Internet/FSE_MEDIA/stelprdb1046486.png

Van Helmond, A. T. M., Mortensen, L. O., Plet-Hansen, K. S., Ulrich, C., Needle, C. L., & Oesterwind, D. (2019). Electronic monitoring in fisheries: Lessons from global experiences and future opportunities. *Fish and Fisheries, 21*(1), 162–189. https://doi.org/10.1111/faf.12425

Van Wijk, S. J., Taylor, M. I., Creer, S., Dreyer, C., & Rodrigues, F. M. (2013). Experimental harvesting of fish populations drives genetically based shifts in body size and maturation. *Frontiers in Ecology and the Environment, 11*(4). https://doi.org/10.1890/120229

Wadea, E., Spalding, A. K., & Biedenweg, K. (2019). Integrating property rights into fisheries management: The case of Belize's journey to managed access. *Marine Policy*, (108).

Wang, W., Liu, H., Li, Y. Q., & Su, J. L. (2014). Development and management of land reclamation in China. *Ocean and Coastal Management, 102*, 415–425.

Wang, Y., Yao, Y., & Ju, M. (2008). Wise use of wetlands: Current state of protection and utilization of chinese wetlands and recommendations for improvement. *Environmental Management, 41*(6), 793–808.

Wang, Y., & Yu, S. (2013). Patterns of terrestrial anthropogenic impacts on coastal wetlands in three city clusters in China. *Acta Ecologica Sinica, 33*(3), 998–1010. (王毅杰, 俞慎 (2013). "三大沿海城市群滨海湿地的陆源人类活动影响模式." 生态学报 *33*(03), 998–1010. In Chinese with English abstract).

Waters, J. R. (2001). Quota management in the commercial red snapper fishery. *Marine Resource Economics, 16*(1), 65–78. https://doi.org/10.1086/mre.16.1.42629314

Waycott, M., Duarte, C. M., Carruthers, T. J. B., Orth, R. J., Dennison, W. C., Olyarnik, S., Calladine, A., Fourquearan, J. W., Heck, K. L., Jr., Hughes, A. R., Kendrick, G. A., Kenworthy, W. J., Short, F. T., & Williams, S. L. (2009). Accelerating loss of seagrasses across the globe threatens coastal ecosystems. *Proceedings of the National Academy of Sciences of the United States of America, 106*(30), 12377–12381. https://doi.org/10.1073/pnas.0905620106

Werner, K., Taylor, M., Diekmann, R., Lloret, J., Möllmann, C., Primicerio, R., & Fock, H. (2019). Evidence for limited adaptive responsiveness to large-scale spatial variation of habitat quality. *Marine Ecology Progress Series, 629*, 179–191. https://doi.org/10.3354/meps13120

Willison, J. H. M., & Côté, R. P. (2009). Counting biodiversity waste in industrial eco-efficiency: Fisheries case study. *Journal of Cleaner Production, 17*(3), 348–353. https://doi.org/10.1016/j.jclepro.2008.08.003

Winther, J. (2018). *Identifying particularly valuable and vulnerable areas* [PowerPoint slides]. Centre for the Ocean and the Arctic & Norwegian Polar Institute.

World Bank, Data Bank. (2018). *Population growth (annual %), China* [Data file]. Retrieved from https://data.worldbank.org/indicator/SP.POP.GROW

World Bank, Food and Agriculture Organization of the United Nations. (2009). *The sunken billions: The economic justification for fisheries reform.* World Bank.

World Bank, Food and Agriculture Organization and WorldFish Center (WB/FAO/WFC). (2012). *Hidden harvest: The global contribution of capture fisheries* (Report No. 66469-GLB). World Bank.

Worm, B., Hilborn, R., Baum, J., Branch, T., Collie, J., Costello, C., & Fogarty, M. (2009). Rebuilding global fisheries. *Science, 325*(5940), 578–585.

Xu, D., & Zhang, G. (2007). Impact of human activities on Coastal Wetlands in China. *Wetland Science, 5*(3), 282–288.

Xu, S. J., Xu, Y. H., Huang, Y. L., & Zheng, F. (2012). Women's roles in the construction of new fishing villages in China, as shown from surveys in Zhejiang Province. *Asian Fisheries Science, 25S*, 229–236.

Xu, Y., Zhang, T., & Zhou, J. (2019). Historical occurrence of Algal Blooms in the Northern Beibu Gulf of China and implications for future trends. *Frontiers in Microbiology, 10*, 451. https://doi.org/10.3389/fmicb.2019.00451

Xuan, J. L., He, Y. Q., Zhou, F., Tang, C., Zheng, X. Y., Liu, H., Yu, L. J., & Chen, J. F. (2019). Aquaculture-induced boundary circulation and its impact on coastal frontal circulation. *Environmental Research Communications, 1*, 051001.

Yang, H. S. (2016). Construction of marine ranching in China: Reviews and prospects. *Journal of Fisheries of China, 40*(7), 1133–1140. (杨红生. 2016. 我国海洋牧场建设回顾与展望. 水产学报, *40*(7), 1133–1140. In Chinese with English abstract).

Yang, H. S., Zhang, S. Y., Zhang, X. M., Chen, P. M., Tian, T., & Zhang, T. (2019). Strategic thinking on the construction of modern marine ranching in China. *Journal of Fisheries of China, 43*(4), 1255–1262. (杨红生, 章守宇, 张秀梅, 陈丕茂, 田涛, 张涛. 2019. 中国现代化海洋牧场建设的战略思考. 水产学报, *43*(4), 1255–1262. In Chinese with English abstract).

Yang, Z. Y. (2005). *Study on the transaction cost of ITQ system in marine fisheries resource management* (Doctoral dissertation, Fudan University, Shanghai, China). (杨正勇, 2005. 海洋渔业资源管理中ITQ制度交易成本研究. 复旦大学博士论文. In Chinese with English abstract).

Yu, R., Zhang, Q., Kong, F., Zhou, Z., Chen, Z., Zhao, Y., Geng, H., Dai, L., Yan, T., & Zhou, M. (2017). Status, impacts and long-term changes of harmful algal blooms in the sea area adjacent to the Changjiang River estuary. *Oceanologia et Limnologia Sinica, 48*, 1178–1186 (于仁成, 张清春, 孔凡洲, 周正熙, 陈振帆, 赵越, 耿慧霞, 代丽, 颜天, 周名江, 2017. 长江口及其邻近海域有害藻华的发生情况、危害效应与演变趋势。海洋与湖沼, *48*(6), 1178–1186. In Chinese). https://doi.org/10.11693/hyhz201709002

Yue, D. D., Wang, L. M., Ji, W. W., Ruan, W., Wang, Q., Xiong, M. S., & Xao, L. (2017). On the relationship between per capita net income of rural residents and per capita consumption of aquatic products in China based on the VAR model. *Fishery Information and Strategy, 32*(3), 161–167. (岳冬冬, 王鲁民, 纪炜炜, 阮雯, 王茜, 熊敏思, 肖黎. 2017. 郑亮基于协整理论和 VAR 模型的中国农村居民人均纯收入与水产品消费关系研究. 渔业信息与战略, *32*(3), 161–167. In Chinese with English abstract).

Zeng, J. N. (2019, April 25). Theoretical study on marine protected area construction in the perspective of sustainable development. *China Natural Resource Daily*, p. 003. (曾江宁, 2019. 可持续发展视野下的海洋保护区建设理论研究. 中国自然资源报, 4 月25 日第003 版. In Chinese).

Zhai, W. D. (2018). Exploring seasonal acidification in the Yellow Sea. *Science China Earth Sciences, 61*. https://doi.org/10.1007/s11430-017-9151-4

Zhai, W. D., Zhao, H. D., Su, J. L., Liu, P. F., Li, Y. W., & Zheng, N. (2019). Emergence of summertime hypoxia and concurrent carbonate mineral suppression in the Central Bohai Sea, China. *Journal of Geophysical Research: Biogeosciences, 124*, 2768–2785.

Zhang, H., & Wu, F. (2017). China's marine fishery and global ocean governance. *Global Policy, 8*, 216–226.

Zhang, H. Z. (2015). China's fishing industry: Current status, government policies, and future prospects. In *China as a "Maritime Power" Conference. CNA Analysis & Solutions*. Arlington, Virginia, USA. Retrieved from https://www.cna.org/cna_files/pdf/China-Fishing-Industry.pdf

Zhang, Q. H., Cheng, J. H., & Xu, H. X. (2007). *Sustainable use of living marine resource in the East China Sea*. Fudan University Press. (张秋华, 程家骅,徐汉祥, 等. 2007. 东海区渔业资源及其可持续利用. 上海:复旦大学出版社. In Chinese).

Zhang, W., Liu, M., Sadovy de Mithcheson, Y., Cao, L., Ledbitter, D., Newton, R., Little, D. C., Li, S., Yang, Y., Chen, X., & Zhou, W. (2019). Fishing for feed in China: Facts, impacts, and implications. *Fish and Fisheries*. https://doi.org/10.1111/faf.12414

Zhang, Y., & Zhao, L. (2018). A comparative study on consumption demand of Chinese and American aquatic products and its implications. *Journal of Ocean University of China—Social Sciences, 5*, 77–84. (张瑛, 赵露. 2018. 中美水产品消费需求对比研究及其启示. 中国海洋大学学报-社会科学版, *5*, 77–84. In Chinese with English abstract).

Zhejiang Government Website. (2014). (浙江政务网, 2014. 水产养殖禁用药物名录) Retrieved August 8, 2019, from http://tazlh.zjzwfw.gov.cn/art/2014/5/30/art_40304_24148.html

Zheng, F., Qiu, G., Fan, H., & Zhang, W. (2013). Diversity, distribution and conservation of Chinese seagrass species. *Biodiversity Science, 21*(5), 517–526. (郑凤英, 邱广龙, 范航清, 张伟 (2013). "中国海草的多样性、 分布及保护." 生物多样性 *21*(05), 517–526. In Chinese with English abstract).

Zheng, Q., Zhang, R., Wang, Y., Pan, X., Tang, J., & Zhang, G. (2012). Distribution of antibiotics in the Beibu Gulf, China: Impacts of river discharge and aquaculture activities. *Marine Environmental Research, 78*, 26–33.

Zhou, Y., Tian, B., Huang, Y., Wu, W., Qi, X., Shu, M., Xu, W., Ge, F., Wei, W., Huang, G., & Zhang, T. (2016). Degradation of coastal wetland ecosystem in China: Drivers, impacts, and strategies. *Bulletin of the Chinese Academy of Sciences, 31*(10), 1157–1166. (周云轩, 田波, 黄颖, 吴文挺, 戚纤云, 舒敏彦, 胥为, 葛芳, 魏伟, 黄盖先, 张婷 (2016). "我国海岸带湿地生态系统退化成因及其对策." 中国科学院院刊 *31*(10), 1157–1166. In Chinese with English abstract).

Zhu, J. Z., & Shi, Y. Y. (2009). Policy analysis on conversion of fishermen in Beibu Gulf. *Pacific Journal*, (8), 77–82. (朱坚真, 师银燕. 2009. 北部湾渔民转产转业的政策分析. 太平洋学报, (8), 77–82. In Chinese).

Zuo, P., Zhao, S., Liu, C., Wang, C., & Liang, Y. (2012). Distribution of Spartina spp. along China's coast. *Ecological Engineering, 40*, 160–166.

The manufacturer's authorised representative in the EU is Springer Nature Customer Service Centre GmbH, Europaplatz 3, 69115 Heidelberg, Germany. If you have any concerns regarding our products, please contact ProductSafety@springernature.com

Printed and bound by CPI Group (UK) Ltd, Croydon, CR0 4YY
01/10/2026
02225922-0020